Economic Modeling Using Artificial Intelligence Methods

Advanced Information and Knowledge Processing

Series Editors
Professor Lakhmi Jain
lakhmi.jain@unisa.edu.au

Professor Xindong Wu
xwu@cs.uvm.edu

For further volumes:
http://www.springer.com/series/4738

Tshilidzi Marwala

Economic Modeling Using Artificial Intelligence Methods

Tshilidzi Marwala
Faculty of Engineering and the Built
    Environment
University of Johannesburg
Johannesburg
South Africa

ISSN 1610-3947
ISBN 978-1-4471-5919-3   ISBN 978-1-4471-5010-7 (eBook)
DOI 10.1007/978-1-4471-5010-7
Springer London Heidelberg New York Dordrecht

Foreword

The quest to find an economic theory that is able to explain all economic activities has alluded both theorists and practitioners alike. If such a theory can be discussed it will usher a new area where a desired economic outcome can be engineered and the social dividends that can be harnessed from this are substantial. However, despite this limitation it has been quite possible to model aspects of the economy separately. For example, it is quite possible to model inflation and also engineer concepts such as inflation targeting.

Economic Modeling Using Artificial Intelligence Methods introduces the concepts of artificial intelligence for modeling economic data. Artificial inteligence is a scientific technique that looks at how nature operates and emulate it. For example, how a human mind works or how a group of ants tackle problems. Artificial intelligence is used particularly in areas where it is incredibly difficult to model a phenomenon.

The recent economic crisis has highlighted the need to treat the area of economic instruments with caution. For example complex derivatives have been responsible for the collapse of banks in the USA. This book deals with the area of understanding economic data, modeling options, understanding economic growth, understanding inflation, controlling inflation, optimizing a portfolio of investment assets and modelling stock market. It also treats the area of interstate peace in promoting economic activities.

This book gives a different perspective of econometrics.

Johannesburg, South Africa   Adam Habib, Ph.D.
March 2013

Preface

Economic Modeling Using Artificial Intelligence Methods introduces the concepts of artificial intelligence for modeling economic data. In this book, the artificial intelligence techniques that are used to model economic data include neural networks, support vector machines, rough sets, genetic algorithm, particle swarm optimization, simulated annealing, multi-agent system, incremental learning and fuzzy networks. In addition, this book explores signal processing techniques to analyze economic data and to deal with vital subjects such as stationarity. These signal processing techniques are the time domain methods, time-frequency domain methods and fractals dimension approaches.

These techniques are used to solve interesting economic problems such as causality versus correlation, modeling the stock market, modeling inflation and portfolio optimization. In addition, game theoretic framework is used to simulate the stock market and control systems technique is used for inflation targeting. Finally, an important area of the relationship between economic dependency and interstate conflict is explored and some interesting insights on how economics can be used to foster peace and vice versa are explored. This book specifically addresses the issue of causality in the non-linear domain and applies the automatic relevance determination, the evidence framework, Bayesian approach and Granger causality to achieve this goal.

This book makes an important contribution to the area of econometrics, and is an interesting read for graduate students, researchers and financial practitioners.

University of Johannesburg   Tshilidzi Marwala
Johannesburg, South Africa
March 2013

Acknowledgements

I would like to thank the following former and current graduate students for their assistance in developing this manuscript: Michael Pires, Msizi Khoza, Ishmael Sibusiso Msiza, Nadim Mohamed, Dr. Brain Leke, Dr. Sizwe Dhlamini, Thando Tettey, Bodie Crossingham, Prof. Fulufhelo Nelwamondo, Vukosi Marivate, Dr. Shakir Mohamed, Dr. Bo Xing, Dr. David Starfield, Dr. Pretesh Patel, Dr. Dalton Lunga, Dr. Linda Mthembu, Dr. Meir Perez, Dr. Megan Russell and Dr. Busisiwe Vilakazi. I thank Mr Evan Hurwitz for developing the ideas in Chap. 9. I also thank colleagues and practitioners that have collaborated directly and indirectly to the writing of the manuscript. In particular, I thank Dr. Ian Kennedy, Dr. Monica Lagazio and the anonymous reviewers for their comments and careful reading of the book. I thank my supervisors Dr. Hugh Hunt, Prof. Stephan Heyns and Prof. Philippe de Wilde.

I dedicate this book to Dr. Jabulile Manana as well as my sons Lwazi Thendo and Nhlonipho Khathutshelo.

University of Johannesburg   Tshilidzi Marwala
Johannesburg, South Africa
March 2013

Contents

1 Introduction to Economic Modeling 1
  1.1   Introduction 1
  1.2   Economic Concepts 2
        1.2.1   Stock Market 2
        1.2.2   Options and Derivatives 2
        1.2.3   Economic Development 3
        1.2.4   Industrialization 4
        1.2.5   Political Stability 4
  1.3   Econometrics 5
        1.3.1   Linear Assumptions 6
        1.3.2   Static Models 6
        1.3.3   Causality Versus Correlation 6
  1.4   Artificial Intelligence 7
        1.4.1   Neural Networks 7
        1.4.2   Support Vector Machines 7
        1.4.3   Autoassociative Networks 8
        1.4.4   Rough Sets 8
        1.4.5   Incremental Learning 8
        1.4.6   Multi-agent Systems 9
        1.4.7   Genetic Algorithms 9
        1.4.8   Particle Swarm Optimization 10
        1.4.9   Control Systems 10
  1.5   Common Mistakes in Data Modeling 11
        1.5.1   Insufficient Datasets 11
        1.5.2   Inappropriate Scaling 12
        1.5.3   Time-Series Tracking 13
        1.5.4   Inappropriate Measures of Performance 13
  1.6   Data Handling 14
        1.6.1   Time Domain Analysis 14
        1.6.2   Frequency Domain 14
        1.6.3   Time-Frequency Domain 15

1.7   Outline of the Book   15
1.8   Conclusions   16
References   17

2   Techniques for Economic Modeling: Unlocking
     the Character of Data   23
2.1   Introduction   23
2.2   Time Domain Data   24
       2.2.1   Average   24
       2.2.2   Variance   26
       2.2.3   Kurtosis   27
2.3   Frequency Doman   28
2.4   Time-Frequency Domain   31
2.5   Fractals   32
       2.5.1   The Rescaled Range (R/S) Methodology   33
       2.5.2   The Hurst Interpretation   35
2.6   Stationarity   36
2.7   Conclusions   40
References   40

3   Automatic Relevance Determination in Economic Modeling   45
3.1   Introduction   45
3.2   Mathematical Framework   46
       3.2.1   Neural Networks   47
       3.2.2   Bayesian Framework   54
       3.2.3   Automatic Relevance Determination   57
3.3   Applications of ARD in Inflation Modeling   58
3.4   Conclusions   60
References   61

4   Neural Approaches to Economic Modeling   65
4.1   Introduction   65
4.2   Multi-layer Perceptron Neural Networks   67
4.3   Radial-Basis Function (RBF)   68
       4.3.1   Model Selection   71
4.4   Support Vector Regression   72
4.5   Applications of MLP, RBF and SVM to Economic Modeling   75
4.6   Conclusion   78
References   78

5   Bayesian Support Vector Machines for Economic
     Modeling: Application to Option Pricing   83
5.1   Introduction   83
5.2   Black-Scholes Model   85
5.3   Bayesian Neural Networks   87
5.4   Hybrid Monte Carlo (HMC)   89

	5.5	Bayesian Support Vector Machines	92
		5.5.1 Monte Carlo Method	94
		5.5.2 Markov Chain Monte Carlo Method.......................	95
	5.6	Experimental Investigation	95
	5.7	Conclusion..	97
	References..	97

6	**Rough Sets Approach to Economic Modeling: Unlocking**	
	Knowledge in Financial Data..........................	**101**
	6.1	Introduction..	101
	6.2	Rough Sets..	103
		6.2.1 Information System...	104
		6.2.2 The Indiscernibility Relation.............................	104
		6.2.3 Information Table and Data Representation...............	105
		6.2.4 Decision Rules Induction..................................	105
		6.2.5 The Lower and Upper Approximation of Sets	105
		6.2.6 Set Approximation	107
		6.2.7 The Reduct...	107
		6.2.8 Boundary Region	108
		6.2.9 Rough Membership Functions	108
	6.3	Discretization Methods	109
		6.3.1 Equal-Frequency-Bin (EFB) Partitioning	109
		6.3.2 Boolean Reasoning	109
		6.3.3 Entropy Based Discretization	110
		6.3.4 Naïve Algorithm...	110
	6.4	Rough Set Formulation	111
	6.5	Application to Modeling the Stock Market	112
	6.6	Conclusion..	115
	References..	115

7	**Missing Data Approaches to Economic Modeling:**	
	Optimization Approach...	**119**
	7.1	Introduction..	119
	7.2	Missing Data Estimation Method	120
	7.3	Auto-associative Networks for Missing Data Estimation	122
	7.4	Particle Swarm Optimization	123
	7.5	Genetic Algorithms (GA)...	125
		7.5.1 Initialization..	127
		7.5.2 Crossover..	127
		7.5.3 Mutation	128
		7.5.4 Selection	128
		7.5.5 Termination	129
	7.6	Simulated Annealing (SA)..	129
		7.6.1 Simulated Annealing Parameters	130
		7.6.2 Transition Probabilities....................................	130

	7.6.3	Monte Carlo Method	130
	7.6.4	Markov Chain Monte Carlo (MCMC)	131
	7.6.5	Acceptance Probability Function: Metropolis Algorithm	131
	7.6.6	Cooling Schedule	132
7.7	Experimental Investigations and Results		132
7.8	Conclusion		133
References			133

8 Correlations Versus Causality Approaches to Economic Modeling... 137
8.1	Introduction	137
8.2	Causality Approach to Economic Modeling	138
8.3	Correlation Machines for Economic Modeling	139
8.4	Classification: Correlation and Causal Machine	143
8.5	Causality	145
8.6	Automatic Relevance Determination (ARD) for Causality	146
8.7	Conclusions	151
References		151

9 Evolutionary Approaches to Computational Economics: Application to Portfolio Optimization ... 155
9.1	Introduction		155
9.2	Background		157
	9.2.1	Modern Portfolio Theory	157
	9.2.2	CAPM Modeling	158
	9.2.3	Genetic Algorithms	159
9.3	Problem Statement		163
9.4	Genetic Algorithm Setup		164
9.5	Analysis		165
	9.5.1	Technical Specifications	165
	9.5.2	Results Analysis	166
9.6	Conclusions		168
References			169

10 Real-Time Approaches to Computational Economics: Self Adaptive Economic Systems ... 173
10.1	Introduction		173
10.2	Incremental Learning		174
10.3	Ensemble Methods		175
	10.3.1	Bagging	176
	10.3.2	Stacking	176
	10.3.3	Adaptive Boosting (AdaBoost)	176
10.4	The Real-Time Method		178
	10.4.1	Learn++ Incremental Learning Method	182
	10.4.2	Confidence Measurement	184
	10.4.3	Multi-layer Perceptron	185

      10.5   Experimental Investigation 186
      10.6   Conclusions 189
      References 190

11 **Multi-agent Approaches to Economic Modeling: Game**
     Theory, Ensembles, Evolution and the Stock Market 195
      11.1   Introduction 195
      11.2   Game Theory 196
      11.3   Multi-agent Systems 197
      11.4   Neural Networks 199
      11.5   Ensembles of Networks 201
           11.5.1 Bagging 201
           11.5.2 Boosting 201
           11.5.3 Stacking 202
           11.5.4 Evolutionary Committees 202
      11.6   Genetic Algorithms 206
      11.7   Simulating the Stock Marketing 207
      11.8   Conclusions 209
      References 210

12 **Control Approaches to Economic Modeling: Application**
     to Inflation Targeting 215
      12.1   Introduction 215
      12.2   Is Inflation Non-stationary? 217
      12.3   Control of Non-stationary Process 219
      12.4   Modeling Inflation 222
      12.5   Controlling Inflation 223
      12.6   Experimental Investigation 227
      12.7   Conclusions 227
      References 228

13 **Modeling Interstate Conflict: The Role of Economic**
     Interdependency for Maintaining Peace 233
      13.1   Introduction 233
      13.2   The Drivers of Interstate Conflict 235
      13.3   Artificial Intelligence 236
           13.3.1 Support Vector Machines (SVMs) for
                   Classifying Conflicts 237
           13.3.2 Fuzzy Sets for Classifying Conflicts 240
           13.3.3 Neuro-Rough Sets for Classifying Conflicts 243
           13.3.4 Automatic Relevance Determination (ARD) 245
      13.4   Controlling Interstate Conflict 246
      13.5   Investigation and Results 247
      13.6   Conclusions 249
      References 250

14 Conclusions and Further Work 253
    14.1   Conclusions 253
    14.2   Further Work 255
    References 255

Biography 257

Index 259

Chapter 1
Introduction to Economic Modeling

Abstract This chapter introduces economic modeling based on artificial intelligence techniques. It introduces issues such as economic data modeling and knowledge discovery, including data mining and causality versus correlation. It also outlines some of the common errors in economic modeling with regard to data handling, modeling, and data interpretation. It surveys the relevant econometric methods and motivates for the use of artificial intelligence methods.

1.1 Introduction

This chapter introduces the topic of economic modeling (Baumol and Blinder 1982; Caldwell 1994; Holcombe 1989; Lange 1945; de Marchi and Blaug 1991). In this book, modeling is defined as the process of creating mathematical and conceptual frameworks for describing economic phenomena. In other words, the outcome of a modeling process, as defined in this book, is a conceptual or mathematical framework that describes how various concepts in economics actually work. The mechanisms, whether mathematical or conceptual, adopted in this book are based on an artificial intelligence framework. Artificial intelligence has been successfully applied to problems such as missing data estimation (Marwala 2009), engineering (Marwala 2010), political science (Marwala and Lagazio 2011a) and condition monitoring (Marwala 2012).

In this book, we define artificial intelligence techniques as mathematical or conceptual processes that are inspired by how nature works. For example, in describing how the gross domestic product is influenced by variables such as average educational levels and international trade volume, we could use a neural network which is based on how a human brain works, to construct a mathematical model that will relate these variables to the gross domestic product. This book will follow this line of thinking, of applying artificial intelligence methods, to describe how various aspects of the economy actually work.

T. Marwala, *Economic Modeling Using Artificial Intelligence Methods*, Advanced Information and Knowledge Processing, DOI 10.1007/978-1-4471-5010-7_1,
© Springer-Verlag London 2013

1.2   Economic Concepts

In this section we describe various economic matters that are addressed in this book, and these include the stock market, options, derivatives, industrialization, economic development and political stability.

1.2.1   Stock Market

One important component of the economy that is considered in this chapter is the stock market (Hamilton 1922; Preda 2009). When companies are listed in the stock market, their net worth is calculated and part of the company is offered to the public to buy shares of that company. This instrument of listing a company in the stock market and thus allowing the public to buy shares in the company is vital for a company to be able to raise financial capital. One additional element that comes about as a result of publicly trading shares is that the price of the stock can end up not reflecting the intrinsic value of the shares. This may result in the over-pricing or underpricing of stock as a result of the lack of knowledge of the real value of stock. For a trader in the stock market, whose primary objective is to maximize financial returns, it is important for there to be instruments that would enable the trader to be able to predict the future price of stocks. In this book, we apply artificial intelligence to predict the future prices of stocks. Applying artificial intelligence techniques for stock market prediction has been conducted quite extensively in the past by practitioners such as Lunga and Marwala (2006), Leke and Marwala (2005) as well as Khoza and Marwala (2011).

1.2.2   Options and Derivatives

As described by Pires and Marwala (2004, 2005), many corporations and companies are exposed to risk in many ways. If a firm's business model is based on exports, then it is exposed to the volatility of the exchange rate. For instance, a diamond mining company is exposed to risk from the diamond price because if the diamond price drops then the mining concern can lose money. Corporations seek to protect themselves from this risk and so what they normally do is that they enter into an agreement to sell diamond at a particular fixed exchange rate and diamond price for the future months. This contract is fixed and the company will neither make nor lose any extra money. The two contracts, the company takes, are known as futures contracts. Because the contract doesn't permit the owner to require any additional money if diamond price increases or the exchange rate weakens, then the company doesn't pay a premium for the contracts. This reduction of risk is called hedging (Ross et al. 2001).

Another mechanism in which companies hedge against this risk is by acquiring options. An option is the right, not the obligation, to buy or sell an underlying asset at a later date, which is called maturity date, by fixing the price of the asset at the present time (Hull 2003). An option that affords the owner the right to buy the asset is called a call option and an option that affords the owner the right to sell the asset is called a put option.

There are two types of options and these are European and American styled options. European options are exercised on the maturity date and American options are exercised on any date leading up to the maturity date. In the example above, with the diamond mining company, the company could buy a commodity put option, permitting the company to sell diamond at a particular price at a fixed time and purchase an exchange rate call option permitting the company to trade at a particular exchange rate at a later date. Options differ from futures in that the owner of the option is offered the right and not the obligation to exercise and thus make them valuable and so companies can benefit from favourable situations in the market and still protect themselves from unattractive effects in the market. The difference between futures and options is that if an unwanted state of affairs happens, then the owner loses the premium that was paid to buy the option. Because of this reason, options are acquired at a premium and there is difficulty identifying the value of this premium. Black and Scholes (1973) formulated a model for pricing options but the difficulty with their model was that it was only applicable for pricing European options (Hull 2003). American options are more difficult to price (Hull 2003) because there is a second random process in the contract because it is not known when the option will be exercised and thus offers the owner of the option an extra level of flexibility (Jarrow and Turnbull 2000).

1.2.3 Economic Development

In this chapter we apply the concept of a developmental state to understand economic development. In order to understand the concept of a developmental state, it is important to highlight some of the characteristics of a developmental state (Thompson 1996; Woo-Cumings 1999). Developmental states generally put strong emphasis on technical education and the development of numeracy and computer skills within the population. This technically oriented education is strategically used to capacitate government structures particularly the bureaucracy. What emerges out of this strategy is that the political and bureaucratic layers are populated by extremely educated people who have sufficient tools of analysis to be able to take leadership initiatives, based on sound scientific basis, at every level of decision making nodes within the government structure. Developmental states have been observed to be able to efficiently distribute and allocate resources and, therefore, invest optimally in critical areas that are the basis of industrialization such as education. The other characteristic that has been observed in successful developmental states is economic nationalism and emphasis on market share over profit. They protect their embryonic domestic industries and focus on aggressive acquisition of foreign technology. This they achieve by deploying their most talented students to overseas universities located in strategic and major centres of the innovation world and also by effectively utilizing their foreign missions (Marwala 2006). Furthermore, they encourage and reward foreign companies that invest in building productive capacity such as manufacturing plants with the aim that the local industrial sector will in time be able to learn vital success factors from these companies. On constructing a harmonious social-industrial complex, developmental states strike a strategic alliance between the state, labour and industry in order to increase critical measures such as productivity, job security and industrial expansion. Even though developmental states do not create enemies unnecessarily and do not participate in the unnecessary criticism of countries with strategic technologies that they would like to acquire, they are, however, sceptical of copying foreign values without translating and infusing them with local characteristics.

1.2.4   Industrialization

The objective of a developmentally oriented state is to create a society in which the citizens are intellectually, socially, economically, and politically empowered. In order to accomplish this objective, certain conditions need to be in place to mobilize social, economic, and political forces to capacitate the state to stimulate the productive forces that would ensure that this goal is achieved. One viewpoint concerning the instrument through which these productive forces can be galvanized is to reorient the country such that adequate productive forces are unleashed to advance industrialization (Marwala 2005a, b; Xing and Marwala 2011).

One critical aspect of industrialization is manufacturing. Building a manufacturing base in a country requires many factors to be in place such as a large number of engineers as well as access to minerals such as copper. The goal of industrialization is to create a country that produces goods and services with high added values. For example, instead of exporting minerals unprocessed, people can be employed to beneficiate these minerals and manufacture goods such as watches and thus add economic value to the final products. The process by which countries add aggregate economic values to the products and services they offer is directly dependant on the level of industrialization in the country's economy.

1.2.5   Political Stability

One important aspect of building a vital economy in a country is political stability, which is characterized by the absence of conflict. One important characteristic of a progressive society is a society which is situated within a state that is at peace with itself, its neighbors, and the international community. Consequently, a democratic society as a matter of principle should aspire for global peace and inspire a culture of the highest form of human development (Marwala and Lagazio 2011a).

Granted that peace is a necessary condition to construct an economically stable society it is, consequently, important to comprehend the anatomy of militarized interstate conflicts and use this understanding to build peaceful, stable, and economically prosperous states in a peaceful and stable international context. The ability to scientifically understand the causes of militarized interstate conflict and then to apply this knowledge to build and spread peace in the international context is indisputably an essential initiative. This book proposes an artificial intelligence perspective to unpack some of the complex behaviours that interstate conflicts display in order to understand the fundamental drivers of war and, therefore, detect early instances of tensions in the international relations arena and, thereby, maximize economic activities (Marwala and Lagazio 2011b).

When a business concern intends to invest in a country, the first criterion to investigate is the stability of a country. Militarized interstate conflict is defined as a threat of military conflict by a country on another (Marwala and Lagazio 2011b). This phenomenon has been modeled extensively and quite successfully using artificial intelligence in the past. Tettey and Marwala (2006a) successfully applied a neuro-fuzzy system for conflict management whereas Habtemariam et al. (2005) successfully applied support vector machines for modeling and managing militarized interstate conflict. Marwala and Lagazio (2004) successfully applied the multi-layered neural networks and genetic algorithms to model and then control militarized interstate disputes, while Tettey and Marwala (2006b) applied a neuro-fuzzy system and genetic algorithms for modeling and controlling militarized interstate disputes. On modeling militarized interstate disputes, it is often important to extract information from the observed data in a linguistic fashion so that these can be used for policy formulation. Tettey and Marwala (2007) applied a neuro-fuzzy system to extract such information from a conflict dataset, while Crossingham et al. (2008) applied optimized rough sets and standard rough sets to extract information from an interstate conflict database.

1.3 Econometrics

Econometrics is a field that applies mathematics and statistics to study economics (Lamy 2012; Spanos 2012; Baldauf and Santos Silva 2012). This is essentially achieved by building mathematical models to explain economic phenomena. Suppose we would like to build a model that relates *car sales* to *inflation*: this necessarily implies that there exists a causal relationship between *inflation* and *car sales*. Mathematically speaking, this relationship may be written as follows:

$$car\ sales = f\ (inflation) \qquad (1.1)$$

The function f can be models such as neural networks or even be linear or linguistic in the case of fuzzy logic. There are three main criticisms that are leveled at traditional econometric modeling and these are described in the next sub-sections.

1.3.1 Linear Assumptions

Traditional econometric models usually assume that relationships are linear, and this assumption is not valid for the majority of the real world cases (Pesaran 1987; Swann 2008). Therefore, it has become essential to build models that are non-linear (Moffitt 1980; Adcock 1995). A neural network has been found to be useful in modeling highly non-linear data because the order of non-linearity is derived or learned directly from the data and not assumed as is the case in many traditional non-linear models.

1.3.2 Static Models

Once the model in Eq. 1.1 is identified, it is generally assumed that the model is valid for all times, which is obviously not the case. It is, therefore, important to build a model that autonomously adapts while it is in use, and traditional econometric techniques have not been very successful in addressing this shortcoming. With the advent of evolutionary programming, it has become quite possible to build models that evolve (Marwala 2005a, b; Nguyen et al. 2012). This book will also address the issue of constructing dynamic models as opposed to building static models within the context of economic modeling.

1.3.3 Causality Versus Correlation

The issue of causality is a problem that has concerned philosophers for many years (Simon and Rescher 1966; Kar et al. 2011; Fallahi 2011). For example, the relationship in Eq. 1.1 inherently assumes causality, that is, the fact that inflation levels influence car sales. However, it might be that this is not the case and what is really the truth is that variables *inflation* and *car sales* are just correlated. This book will address this matter.

This book will deal with the problems of static models, linear assumptions, and causality versus correlation problem by applying artificial intelligence which is described in the next section.

1.4 Artificial Intelligence

This section presents an overview of artificial intelligence techniques which are applied for economic modeling in this book.

1.4.1 Neural Networks

One important type of artificial intelligence techniques is a neural network. Neural networks are computational tools that may be viewed as being inspired by how the brain functions and applying this framework to construct mathematical models. Neural networks estimate functions of arbitrary complexity using given data. Supervised neural networks are used to represent a mapping from an input vector onto an output vector, while unsupervised networks are used to classify the data without prior knowledge of the classes involved. In essence, neural networks can be viewed as generalized regression models that have the ability to model data of arbitrary complexities. There are many types of neural networks and the most common neural network architectures are the multilayer perceptron (MLP) and the radial basis function (RBF) (Bishop 1995). Neural networks have been applied successfully in many different areas of varying complexities. Soares et al. (2006) applied neural networks in flight control, while Shukla et al. (2012) applied neural networks for software maintenance. Xing et al. (2010) applied neural networks for machine clustering while Nelwamondo et al. (2009) applied neural networks and dynamic programming for missing data estimation in biomedical applications. Because of these extensive successes of neural networks, this book will apply neural networks for modelling economic data.

1.4.2 Support Vector Machines

Support vector machines are supervised learning approaches used largely for classification, and originated from statistical learning theory and were first introduced by Vapnik (1998). The use of support vector machines to model complex systems has been a subject of research for many years. Successful implementations of support vector machines to model complicated systems include Marwala et al. (2007) who applied this method for damage detection in structures, Msiza et al. (2007) who applied this method for forecasting the water demand time series, as well as Patel and Marwala (2009) who applied support vector machines for caller behavior classification. Due to these successful applications of support vector machines, this book also applies support vector machines for modelling economic data.

1.4.3 Autoassociative Networks

An auto-associative network is a model that is trained to recall its inputs. These networks are sometimes called auto-encoders or memory networks (Kramer 1992; Fu and Yan 1995; Marseguerra and Zoia 2006; Marwala 2012; Turova 2012). This means that, whenever an input is presented to the network, the output is the predicted input. These networks have been used in a number of applications including novelty detection, feature selection, and data compression. In this book, we propose to use auto-associative networks to construct a missing data estimation technique predictive system, which was described by Marwala (2009). Auto-associative networks have been applied successfully in many diverse areas such as HIV modelling (Leke et al. 2006), damage detection in structures (Zhou et al. 2011), and for predicting internet stability (Marais and Marwala 2004).

1.4.4 Rough Sets

Rough set theory, which was proposed by Pawlak (1991), is a mathematical method which models vagueness and uncertainty. It allows one to approximate sets that are difficult to explain even with accessible information (Marwala 2012). As observed by many researchers in the past, the advantages of rough sets, as with many other artificial intelligence methods, are that they do not require inflexible *a priori* assumptions about the mathematical characteristics of such complex relationships, as generally required for the multivariate statistical methods (Machowski and Marwala 2005; Crossingham et al. 2009; Marwala and Lagazio 2011a). Rough set theory is premised on the assumption that the information of interest is associated with some information from its universe of discourse (Crossingham and Marwala 2007, 2009; Marwala and Crossingham 2008, 2009; Marwala 2012). This technique is useful in that unlike neural networks, for example where the identified model is a strictly mathematical concept, in rough sets method the identified model that describes the data is in terms of natural language. Because of this reason, rough sets are useful in economics because they in a sense represent a formal technique for knowledge extraction from data.

1.4.5 Incremental Learning

Incremental learning methods are approaches that are able to learn incrementally. Incremental learning is suitable for modelling dynamically time-varying systems where the operating conditions change with time (Pang et al. 2012; Clemente et al. 2012; Khreich et al. 2012). It is also suitable when the data set accessible is inadequate and does not completely characterize the system (Marwala

2012; Cavalin et al. 2012; Martínez-Rego et al. 2012; Li et al. 2012a). Another advantage of incremental learning is that it can take into account the new conditions that may be presented by the newly acquired data. For example, suppose a model for predicting inflation is built, the concept of incremental learning can be applied so that every month when new inflation data set comes out, the model is updated without having to reconstruct it entirely. In this book we apply incremental learning techniques that are based on ensemble methods.

Ensemble learning is a method where multiple models are identified and combined to solve a specific problem (Rogova 1994; Polikar 2006). Ensemble learning is usually applied to increase the performance of a model (Zhang et al. 2012a; Li et al. 2012b; Hu et al. 2012; Ñanculef et al. 2012). In this book, ensemble based methods are applied for incremental learning in the modeling of economic data.

1.4.6 Multi-agent Systems

Agents are computer systems that are located in specific environments and are capable of *autonomous action* in this environment with the aim of meeting its objectives (Marivate et al. 2008; Wooldridge 2004; Baig 2012; Hu 2012). Intelligent agents are agents that are capable of *reacting* to changes in their environment and they possess *social ability* such as communication, as well as interaction and the ability to use artificial intelligence to achieve their objectives by being *proactive* (Wooldridge 2004; Rudowsky 2004; Zhang et al. 2012b). Agents are active, modeled to achieve specific tasks, and are able to autonomously act and take decisions (Hu 2012; Chen and Wang 2012). In an object oriented framework, objects are passive, non-oriented and modeled to represent things (Weisfeld 2009; Schach 2006; Abadi and Cardelli 1998). Consequently, the agent modeling method can be viewed as a more powerful manifestation of localized computational units that execute definite tasks while objects model real-world "things" with particular characteristics. A multi-agent system (MAS) is a combination of multiple agents in one system to solve a problem (Baig 2012). These systems have agents that are able to solve problems that are simpler than the total system. They can communicate with one another and support each other in realizing larger and more complex objectives (Hurwitz and Marwala 2007; van Aardt and Marwala 2005). Multi-agent systems have been applied in simulating trading systems (Mariano et al. 2001) and industrial automation (Wagner 2002).

1.4.7 Genetic Algorithms

The genetic algorithm approach is a population-based, probabilistic method that is intended to identify a solution to a problem from a population of possible solutions

(Velascoa et al. 2012; Goyal and Aggarwal 2012). It is inspired by Darwin's theory of natural evolution where the principle of 'the survival of the fittest' applies and members of the population compete to survive and reproduce while the weaker ones disappear from the population (Darwin 1859). Each individual is allocated a fitness value in accordance to how well it achieves the aim of solving the problem. New and more evolutionary fit individual solutions are created during a cycle of generations, where selection and re-combination operations, analogous to gene transfer are applied to the current individuals. This continues until a termination condition is satisfied. Genetic algorithms have been applied successfully in many areas such as engineering (Marwala 2002), control systems (Marwala 2004), and condition monitoring of buildings (Marwala and Chakraverty 2006).

1.4.8 Particle Swarm Optimization

The particle swarm optimization (PSO) method was proposed by Kennedy and Eberhart (1995). This technique was inspired by algorithms that model the "flocking behavior" seen in birds. Researchers in artificial life (Reynolds 1987; Heppner and Grenander 1990) developed simulations of bird flocking. In the context of optimization, the concept of birds finding a roost is analogous to a process of finding an optimal solution. PSO is a stochastic, population-based evolutionary algorithm that is extensively used for the optimization of complex problems (Marwala 2010). It is based on socio-psychological principles that are inspired by swarm intelligence, which gives understanding into social behavior and has contributed to engineering applications. Society enables an individual to influence and learn to solve problems by communicating and interacting with other individuals and, in that way, develop similar approach of solving problems. Thus, swarm intelligence is driven by two factors (Kennedy and Eberhart 1995; Marwala 2009, 2010, 2012):

1. Group knowledge.
2. Individual knowledge.

Each member of a swarm always behaves by balancing between its individual knowledge and the group knowledge.

1.4.9 Control Systems

A control system is essentially a procedure where the input of a system is manipulated to obtain a particular desired outcome (Marwala and Lagazio 2004; Boesack et al. 2010; Zhu et al. 2012). To realize this, a model that describes the relationship between the input and the outcome needs to be identified. For example, suppose we have a model that describes the relationship between the growth domestic product (GDP) and the interest rate to the inflation rate. Such a relationship can be identified using various methods such as neural networks. After this model that predicts inflation rate given the interest rate and GDP has been identified, the following stage is to apply a control system to identify the interest rate that gives the desired inflation rate. This involves an optimization procedure such as genetic algorithms or particle swarm optimization. In economics, this process is called inflation targeting. Control techniques have been applied successfully in many diverse fields such as brewing (Marwala 2004), water supply systems (Wu et al. 2012), and political science (Marwala et al. 2009).

1.5 Common Mistakes in Data Modeling

The prospect of artificial intelligence for application in economic modeling has been apparent to virtually everyone to have even experimented in the fields since their inception. Efforts to make such artificial intelligence applications accepted within the economic modeling space have generally been unsuccessful, and not without justifications. Mistakes have continued to give knowledge and theories that have passed superficial academic enquiry, but have demonstrated to be unsuccessful when applied to actual real world data. As identified by Hurwitz and Marwala (2012), a number of common mistakes will be probed, demonstrating how they frequently slip past cursory academic inquiry, and then displaying how they fail when applied to actual real data and why. It is intended that with attention paid to common shortcomings, researchers can circumvent these drawbacks and improve the uses of these methods within the economic modeling arena. Even though many of these errors overlap in individual implementations, this chapter intends to tackle them on an individual basis in order to more easily circumvent these mistakes in the future.

1.5.1 Insufficient Datasets

According to Hurwitz and Marwala (2012), conventional trading strategies are complicated systems, usually entailing a cycle of prediction, evaluation, feedback, and recalibration when designed. The cycle contains a feedback element but not necessarily the predictive technique. This cycle comprises the designer predicting price movements then evaluating trades based on price movements. The intention for recalibration feedback going into both prediction and evaluation instruments is that in many more complex systems the actual trading based on the predictions will be updated, as well as the predictions themselves. Because of this cycle, it becomes essential to have a strictly unseen set of data to evaluate performance. Normally, trading systems use a training data set and a verification data set (de Oliveira et al. 2011; Lam 2004; Kim 2002; Bao and Yang 2008) and this is not adequate, as neither the predictive system nor the evaluation system ever see the second set of data during the optimization phase. This is nevertheless inadequate, as the calibration of the training system is conducted with the results of the verification set taken into account, necessitating a totally unseen validation set to be used so as to validate if the system is truly generalizing. Overlooking this critical phase can give a system result that has merely been tweaked by the designer to fit the specific data, without essentially being able to function correctly in a general setting. The results will obviously appear satisfactory as the system has been calibrated precisely to fit the data being used even while the designer has not planned this to be the case. In the case that the trading mechanism is revisited after the validation set has been utilized, a new set of unseen validation data must be achieved.

1.5.2 Inappropriate Scaling

As identified by Hurwitz and Marwala (2012), this error is characterized by expressing the normally large target values of the predicted variable as its actual value (Kaastra and Boyd 1996), instead of scaling the data to some appropriate level within (or near) the range of the training data. The justification is to offer a precise understanding of the actual target values.

The actual data and the predicted data will appear to be close in the initial prediction, but experience difficulty to reach the higher values in the range for the later values. The actual errors in this initial prediction will obviously be very low, being underestimated by the scale of the data, despite their being fairly clearly unusable for any undertaking that necessitates the prediction. Actually, the errors characteristic of the system are frequently hidden by using unsuitable measures of accuracy or performance. It is actually far more risky to commit this error if the target data is reasonably bounded, as the evident lack of fit will not be obvious, and what is in effect an unusable prediction can certainly be confused for a performing predictive system, and leads to all the dangers characteristic in trading upon poor information.

The reason for this discrepancy is the high quantitative value of the predictive results, which offer a low registered error for what is really a large trading error. Considering a prediction for a given input–output set with the correct value being 1,025 and the system's predicted value being 1,010. The actual error in root mean squares (RMS) terms is small, while the effect on predictions is actually quite high, considering a daily expected fluctuation of approximately 15 cents, which explains why an error that appears so small is actually significant enough to render a predictive system unusable. The clear recommendation is to first pre-process the data and as part of that process to scale the data. Depending on the nature of the historical share price fluctuations, a scaling factor of anything from the maximum historical price recorded to a fractional amount larger than the maximum historical price can be applied. It is recommended, even though not necessarily relevant to this particular error that one also scales input data for ease of training and convergence.

1.5.3 Time-Series Tracking

This error occurs in time series modeling where the predictive system predicts the previous day's price as the present price which satisfies the error minimization function's requirement (Hurwitz and Marwala 2012). This error emanates when an attempt is made to do an exact price prediction based upon the time-series data of historical prices of the self-same share. Unfortunately, any trading based upon such a system is completely unusable as it cannot ever predict an accurate price movement unless by some coincidence every single day's new close is the same as that of the day before. To avoid this error, it is then necessary for the user to reconsider the input–output pairs for the system to learn from and consider change in prices as an output rather than the price itself. When trading, the direction of price movement is actually far more important than the precise amount, and this distinction is critical if a trading system is to be successful.

1.5.4 Inappropriate Measures of Performance

The problem here lies not with the measurements themselves, but rather on the reliance on them for validating the success of a trading system. These methods often obscure problems in the system design by looking like successful computational intelligence systems by the standard computational measures (Hurwitz and Marwala 2012). This includes graphs of receiver operating characteristics (ROC) curve and other typical computational measures of performance. If any of the preceding errors had been made, they would not be detected by the usual performance measures since they only measure the performance of the system based on the given input and output values. This is a dangerous error to commit, as the system is still concealing any mistakes made, but the user is satisfied to carry on, secure in the success of the system, verified by an inappropriate measure of performance. Dependence on these measures arises naturally to users in this field as they form the benchmark of most computational intelligence and machine learning approaches, and are, therefore, likely to be utilized almost out of practice.

Instead of the above, the user should set up a trading simulator, and apply the designed predictor to simulate trading based on its predictions. By performing actual trades based upon the predictions, many of the errors described will be quickly identified, as the actual trading results will be poor, or at best highly erratic. The nature of the errors will often become apparent when measuring performance in this manner, matching those described within each section, making it a much more useful measure of performance both during and after the system design process.

1.6   Data Handling

There are many ways in which economic data can be handled when used in economic modeling. The choices on how economic data are handled sometimes have effects which artificial intelligence modeling method to be used. In this section, we describe three domains that could be used to model economic data and these are time, frequency and time-frequency domains.

1.6.1   Time Domain Analysis

Time domain data is un-processed data taken over a time history. For example, if we consider GDP data as a function of time, this will be said to be in the time domain. From this data in the time domain, essential statistical features such as means, variance, Kurtosis can be extracted (Marwala 2012). Normally when these data are used, some of the statistical analysis such as variance and means are used. Lima and Xiao (2007) used economic data in the time domain to evaluate whether shocks last forever while Kling and Bessler (1985) compared multivariate forecasting procedures for economic time series analysis. Bittencourt (2012) studied the inflation and economic growth in Latin America in the time domain, while Greasley and Oxley (1998) compared British and American economic and industrial performance between years 1860 and 1993 in the time domain. Even though studying economic phenomena in the time domain is useful, it is sometimes necessary to study economic data in the frequency domain which is described in the next section.

1.6.2   Frequency Domain

The measured economic data in the time domain (time series) can be transformed into the frequency domain using Fourier transforms (Fourier 1822). As an example, the GDP versus time data can be transformed into the frequency domain using the Fourier transform and then this signal can be represented in magnitude and phase versus frequency data. The data in the frequency domain will have a series of peaks and troughs with each peak corresponding to the frequency of each cycle that makes the data. McAdam and Mestre (2008) evaluated macro-economic models in the frequency domain, while Tiwari (2012) conducted an empirical investigation of causality between producers 'price and consumers' price indices in Australia in the frequency domain. Gradojevic (2012) applied frequency domain analysis to study foreign exchange order flows while Grossmann and Orlov (2012) studied the exchange rate misalignments in the frequency domain.

1.6.3 Time-Frequency Domain

Most economic data are highly non-linear and non-stationary signals. None stationary signals are those whose frequency components change as a function of time (Larson 2007; Marwala 2012). To analyze non-stationary signals, the application of the Fast Fourier Transform method is not satisfactory. Consequently, time-frequency methods that concurrently display the time and frequency components of the signals are essential. Some of the time-frequency methods that have been used are: the Short-Time Fourier Transform (STFT), Wavelet Transform (WT) and Wigner-Ville Distribution (WVD). Gallegati (2008) applied wavelet analysis in stock market analysis, while Yogo (2008) applied wavelet analysis for measuring business cycles. Furthermore, Benhmad (2012) applied wavelet analysis for modeling nonlinear Granger causality between the oil price and the U.S. dollar.

1.7 Outline of the Book

In Chap. 2, data modeling techniques in economic modeling are studied. These methods include concepts such as mean, variance and fractals and how these vital concepts are applied to economics. Frequency and time-frequency analysis techniques are also studied.

Chapter 3 introduces the Bayesian and the evidence frameworks to construct an automatic relevance determination method. These techniques are described in detail, relevant literature reviews are conducted and their use is justified. The automatic relevance determination technique is then applied to determine the relevance of economic variables that are essential for driving inflation rate. Conclusions are drawn and are explained within the context of economic sciences.

Chapter 4 describes the multi-layered perceptron, radial basis functions, and support vector machines and apply these to economic modeling. The maximum-likelihood techniques are implemented to train these networks.

Chapter 5 introduces Bayesian support vector machines and multi-layer pereceptron for option pricing. European styled options can be priced using the Black-Scholes equation and are only exercised at the end of the period but American options can be exercised at any time during the period and are, therefore, more complex due to the second random process they introduce. These techniques are implemented using a Bayesian approach to model American options and the results are compared.

Chapter 6 introduces a rough set approach to economic modeling. A rough set theory based predictive model is implemented for the financial markets. The theory can be used to extract a set of reducts and a set of trading rules based on trading data.

Chapter 7 introduces an autoassociative network, with optimization methods, for modeling economic data. The autoassociative network is created using the multi-layered perceptron network while the optimization techniques which are implemented are genetic algorithms, particle swarm optimization, and simulated annealing. The results obtained for modeling inflation are then compared.

Chapter 8 explores the issue of treating a predictive system as a missing data problem, that is, correlation exercise and compares it to treating it as a cause and effect exercise, that is, feed-forward network. An auto-associative neural network is combined with a genetic algorithm and then applied to missing economic data estimation. The results of the missing data imputation approach are compared to those from a feed-forward neural network.

Chapter 9 examines the use of a genetic algorithm in order to perform the task of constantly rebalancing a portfolio targeting specific risk and return characteristics. Results of targeting both the risk and return are investigated and are compared as well as optimizing the non-targeted variable in order to create efficient portfolios.

Chapter 10 introduces real-time approaches to economic modeling. This chapter assumes that a complete model is the one that is able to continuously self-adapt to the changing environment. In this chapter, an incremental algorithm that is created to classify the direction of movement of the stock market is proposed and applied.

Chapter 11 introduces multi-agent approaches within game theoretic framework and applies this to model a sock market. This multi-agent system learns by using neural networks and adapts using genetic programming.

Chapter 12 applies control approaches to economic modeling and applies this to inflation targeting. In this chapter, a control system approach that is based on artificial intelligence is adopted to analyze the inflation targeting strategy.

Chapter 13 explores the role of trade in maintaining peace and, therefore, healthy economic activities. This is done by constructing the relationship between independent variables Allies, Contingency, Distance, Major Power, Capability, Democracy as well as Dependency which indicates inter-country trade and the dependent variable Interstate Conflict.

In Chap. 14 conclusions are drawn and future and emerging areas in economic modeling are identified and emerging opportunities are drawn.

1.8   Conclusions

This chapter introduced economic modeling based on artificial intelligence methods. It introduced issues such as economic data handling and modeling as well as prediction, knowledge discovery including data mining, and causality versus correlation. It also outlined some of the common problems in economic modeling with regards to data handling, modeling, and data interpretation. It surveyed the relevant econometric methods and motivated for the use of artificial intelligence methods.

References

Abadi M, Cardelli L (1998) A theory of objects. Springer, Berlin

Adcock C (1995) Non-linear dynamics chaos and econometrics. Int J Forecast 11:599–601

Baig ZA (2012) Multi-agent systems for protecting critical infrastructures: a survey. J Netw Comput Appl 35:1151–1161

Baldauf M, Santos Silva JMC (2012) On the use of robust regression in econometrics. Econ Lett 114:124–127

Bao D, Yang Z (2008) Intelligent stock trading system by turning point confirming and probabilistic reasoning. Expert Syst Appl 34:620–627

Baumol W, Blinder A (1982) Economics: principles and policy. Harcourt Brace Jovanovich, New York

Benhmad F (2012) Modeling nonlinear Granger causality between the oil price and U.S. dollar: a wavelet based approach. Econ Model 29:1505–1514

Bishop CM (1995) Neural networks for pattern recognition. Oxford University Press, Oxford

Bittencourt M (2012) Inflation and economic growth in Latin America: some panel time-series evidence. Econ Model 29:333–340

Black F, Scholes M (1973) The pricing of options and corporate liabilities. J Politic Econ 81: 637–659

Boesack CD, Marwala T, Nelwamondo FV (2010) Application of GA-fuzzy controller design to automatic generation control. In: Proceedings of the IEEE IWACI, Suzhou, 2010, pp 227–232

Caldwell B (1994) Beyond positivism: economic methodology in the twentieth century. Routledge, New York

Cavalin PR, Sabourin R, Suen CY (2012) LoGID: an adaptive framework combining local and global incremental learning for dynamic selection of ensembles of HMMs. Pattern Recogn 45:3544–3556

Chen YM, Wang B-Y (2012) A study on modeling of human spatial behavior using multi-agent technique. Expert Syst Appl 39:3048–3060

Clemente IA, Heckmann M, Wrede B (2012) Incremental word learning: efficient HMM initialization and large margin discriminative adaptation. Speech Comm 54:1029–1048

Crossingham B, Marwala T (2007) Using optimisation techniques to granulise rough set partitions. Comput Models Life Sci 952:248–257, American Institute of Physics

Crossingham B, Marwala T, Lagazio M (2008) Optimized rough sets for modelling interstate conflict. In: Proceedings of the IEEE international conference on systems, man, and cybernetics, Singapore, pp 1198–1204

Crossingham B, Marwala T, Lagazio M (2009) Evolutionarily optimized rough sets partitions. ICIC Expr Lett 3:241–246

Darwin C (1859) On the origin of species., Chapter XIV. John Murray Publisher, London. ISBN 0-8014-1319-2

de Marchi NB, Blaug M (1991) Appraising economic theories: studies in the methodology of research programs. Edward Elgar Publishing, Brookfield

de Oliveira FA, Zarate LE, de Azevedo Reis M, Nobre CN (2011) The use of artificial neural networks in the analysis and prediction of stock prices. In: Proceedings of the IEEE international conference on systems, man, and cybernetics, Anchorage, 2011, pp 2151–2155

Fallahi F (2011) Causal relationship between energy consumption (EC) and GDP: a Markov-switching (MS) causality. Energy 36:4165–4170

Fourier JB (1822) Théorie Analytique de la Chaleur. Chez Firmin Didot, père et fils, Paris

Fu AMN, Yan H (1995) Distributive properties of main overlap and noise terms in autoassociative memory networks. Neural Nets 8:405–410

Gallegati M (2008) Wavelet analysis of stock returns and aggregate economic activity. Comput Stat Data Anal 52:3061–3074

Goyal MK, Aggarwal A (2012) Composing signatures for misuse intrusion detection system using genetic algorithm in an offline environment. Adv Intell Syst Comput 176:151–157

Gradojevic N (2012) Frequency domain analysis of foreign exchange order flows. Econ Lett 115:73–76

Greasley D, Oxley L (1998) Comparing British and American economic and industrial performance 1860–1993: a time series perspective. Explor Econ Hist 35:171–195

Grossmann A, Orlov AG (2012) Exchange rate misalignments in frequency domain. Int Rev Econ Finance 24:185–199

Habtemariam E, Marwala T, Lagazio M (2005) Artificial intelligence for conflict management. In: Proceedings of the IEEE international joint conference on neural networks, Montreal, 2005, pp 2583–2588

Hamilton WP (1922) The stock market baraometer. Wiley, New York

Heppner F, Grenander U (1990) A stochastic non-linear model for coordinated bird flocks. In: Krasner S (ed) The ubiquity of chaos, 1st edn. AAAS Publications, Washington, DC

Holcombe R (1989) Economic models and methodology. Greenwood Press, New York

Hu G (2012) Robust consensus tracking of a class of second-order multi-agent dynamic systems. Syst Contr Lett 61:134–142

Hu J, Fennel K, Mattern JP, Wilkin J (2012) Data assimilation with a local Ensemble Kalman Filter applied to a three-dimensional biological model of the Middle Atlantic Bight. J Mar Syst 94:145–156

Hull JC (2003) Options, futures and other derivatives. Prentice Hall, Upper Saddle River

Hurwitz E, Marwala T (2007) Multi-agent modeling using intelligent agents in a game of Lerpa. http://arxiv.org/ftp/arxiv/papers/0705/0705.0693.pdf. Last accessed in 7 March 2013

Hurwitz E, Marwala T (2012) Common mistakes when applying computational intelligence and machine learning to stock market modelling. http://arxiv.org/ftp/arxiv/papers/1208/1208.4429.pdf. Last accessed in 7 March 2013

Jarrow RA, Turnbull SM (2000) Derivative securities. South-Western College Publishing, Cambridge, MA

Kaastra L, Boyd M (1996) Designing a neural network for forecasting financial and economic time series. Neurocomputing 10:215–236

Kar M, Nazlıoğlu Ş, Ağır H (2011) Financial development and economic growth nexus in the MENA countries: bootstrap panel granger causality analysis. Econ Model 28:685–693

Kennedy J, Eberhart R (1995) Particle swarm optimization. Proc IEEE Int Conf Neural Netw 4:1942–1948

Khoza M, Marwala T (2011) A rough set theory based predictive model for stock prices. In: Proceedings of the IEEE 12th international symposium on computer intelligent and information, Budapest, Hungary, pp 57–62

Khreich W, Granger E, Miri A, Sabourin R (2012) A survey of techniques for incremental learning of HMM parameters. Info Sci 197:105–130

Kim K (2002) Financial time series forecasting using support vector machines. Neurocomputing 55:307–319

Kling JL, Bessler DA (1985) A comparison of multivariate forecasting procedures for economic time series. Int J Forecast 1:5–24

Kramer MA (1992) Autoassociative neural networks. Comp Chem Eng 16:313–328

Lam M (2004) Neural network techniques for financial performance prediction: integrating fundamental and technical analysis. Decis Support Syst 37:567–581

Lamy L (2012) The econometrics of auctions with asymmetric anonymous bidders. J Econ 167:113–132

Lange OR (1945) The scope and method of economics. Rev Econ Stud 13:19–32

Larson DR (2007) Unitary systems and wavelet sets. In: Wavelet Analysis and Applications by Qian, Tao; Vai, Mang I.; Xu, Yuesheng (Eds.) Birkhäuser, 2007, pp 143–171

Leke B, Marwala T (2005) Optimization of the stock market input time-window using Bayesian neural networks. In: Proceedings of the IEEE international conference on service operations, logistics and informatics, Beijing, 2005, pp 883–894

Leke BB, Marwala T, Tettey T (2006) Autoencoder networks for HIV classification. Curr Sci 91:1467–1473

Li H, Zhang T, Qiu R, Ma L (2012a) Grammar-based semi-supervised incremental learning in automatic speech recognition and labeling. Energy Procedia 17B:1843–1849

Li K, Liu Z, Han Y (2012b) Study of selective ensemble learning methods based on support vector machine. Phys Procedia 33:1518–1525

Lima LR, Xiao Z (2007) Do shocks last forever? Local persistency in economic time series. J Macroecon 29:103–122

Lunga D, Marwala T (2006) Online forecasting of stock market movement direction using the improved incremental algorithm. Lect Notes Comput Sci 4234:440–449

Machowski LA, Marwala T (2005) Using object oriented calculation process framework and neural networks for classification of image shapes. Int J Innovative Comput Info Contr 1:609–623

Marais E, Marwala T (2004) Predicting global internet instability caused by worms using neural networks. In: Proceedings of the annual symposium of the Pattern Recognition Association of South Africa, Cape Town, 2004, pp 81–85

Mariano P, Pereira C, Correira L, Ribeiro R, Abramov V, Szirbik N, Goossenaerts J, Marwala T, de Wilde P (2001) Simulation of a trading multi-agent system. Proc IEEE Int Conf Syst Man Cybern 5:3378–3384

Marivate V, Ssali G, Marwala T (2008) An intelligent multi-agent recommender system for human capacity building. In: Proceedings of the 14th IEEE Mediterranean electrotechnical conference, Ajaccio, 2008, pp 909–915

Marseguerra M, Zoia A (2006) The autoassociative neural network in signal analysis III. Enhancing the reliability of a NN with application to a BWR. Ann Nucl Energy 33:475–489

Martínez-Rego D, Fontenla-Romero O, Alonso-Betanzos A (2012) Nonlinear single layer neural network training algorithm for incremental, nonstationary and distributed learning scenarios. Pattern Recogn 45:4536–4546

Marwala T (2002) Finite element updating using wavelet data and genetic algorithm, Amer Inst of Aeronaut and Astronaut. J Aircraft 39:709–711

Marwala T (2004) Control of complex systems using Bayesian neural networks and genetic algorithm. Int J Eng Simu 5:28–37

Marwala T (2005a) Evolutionary optimization methods in finite element model updating. In: Proceedings of the international modal analysis conference, Orlando, Florida

Marwala T (2005b) Strategies and tactics for increasing economic participation. Umrabulo 24:41–43

Marwala T (2006) Skills necessary for the advancement of South Africa. Umrabulo 26:60–61

Marwala T (2009) Computational intelligence for missing data imputation, estimation and management: knowledge optimization techniques. IGI Global Publications, New York

Marwala T (2010) Finite element model updating using computational intelligence techniques. Springer, London

Marwala T (2012) Condition monitoring using computational intelligence methods. Springer, London

Marwala T, Chakraverty S (2006) Fault classification in structures with incomplete measured data using autoassociative neural networks and genetic algorithm. Curr Sci 90:542–548

Marwala T, Crossingham B (2008) Neuro-rough models for modelling HIV. In: Proceedings of the IEEE international conference on systems, man, and cybernetics, Singapore, 2008, pp 3089–3095

Marwala T, Crossingham B (2009) Bayesian rough sets. ICIC Express Lett 3:115–120

Marwala T, Lagazio M (2004) Modelling and controlling interstate conflict. In: Proceedings of the IEEE international joint conference on neural networks, Budapest, 2004, pp 1233–1238

Marwala T, Lagazio M (2011a) Militarized conflict modeling using computational intelligence techniques. Springer, London

Marwala T, Lagazio M (2011b) The anatomy of interstate conflicts: an artificial intelligence perspective. Think 30:40–42

Marwala T, Mahola U, Chakraverty S (2007) Fault classification in cylinders using multi-layer perceptrons, support vector machines and Gaussian mixture models. Comput Assisted Mech Eng Sci 14:307–316

Marwala T, Lagazio M, Tettey T (2009) An integrated human-computer system for controlling interstate disputes. Int J Comput Appl 31:2402–2410

McAdam P, Mestre R (2008) Evaluating macro-economic models in the frequency domain: a note. Econ Model 25:1137–1143

Moffitt R (1980) Disequilibrium econometrics and non-linear budget constraints. Econ Lett 5:49–52

Msiza IS, Nelwamondo FV, Marwala T (2007) Artificial neural networks and support vector machines for water demand time series forecasting. In: Proceedings of the IEEE international conference on systems, man, and cybernetics, Montreal, 2007, pp 638–643

Ñanculef R, Valle C, Allende H, Moraga C (2012) Training regression ensembles by sequential target correction and resampling. Inf Sci 195:154–174

Nelwamondo FV, Golding D, Marwala T (2009) A dynamic programming approach to missing data estimation using neural networks. Inf Sci. doi:10.1016/j.ins.2009.10.008

Nguyen TT, Yang S, Branke J (2012) Evolutionary dynamic optimization: a survey of the state of the art. Swarm Evol Comput 6:1–24

Pang Y, Deng J, Yuan Y (2012) Incremental threshold learning for classifier selection. Neurocomputing 89:89–95

Patel PB, Marwala T (2009) Genetic algorithms, neural networks, fuzzy inference system, support vector machines for call performance classification. In: Proceedings of the IEEE international conference on machine learning and applications, San Antonio, 2009, pp 415–420

Pawlak Z (1991) Rough sets – theoretical aspects of reasoning about data. Kluwer Academic Publishers, Dordrecht

Pesaran MH (1987) Econometrics. In: The new Palgrave: a dictionary of economics, vol 2. Macmillan/Stockton Press/Maruzen, London/New York/Tokyo, p 8

Pires MM, Marwala T (2004) Option pricing using neural networks and support vector machines. In: Proceedings of the IEEE international conference on systems, man, and cybernetics, The Hague, 2004, pp 1279–1285

Pires MM, Marwala T (2005) American option pricing using Bayesian multi-layer perceptrons and Bayesian support vector machines. In: Proceedings of the IEEE 3rd international conference on computational cybernetis, Mauritius, 2005, pp 219–224

Polikar R (2006) Ensemble based systems in decision making. IEEE Circuits Syst Manag 6:21–45

Preda A (2009) Framing finance: the boundaries of markets and modern capitalism. University of Chicago Press, Chicago

Reynolds CW (1987) Flocks, herds and schools: a distributed behavioral model. Comput Graph 2:25–34

Rogova G (1994) Combining the results of several neural network classifiers. Neural Netw 7:777–781

Ross S, Westerfield RW, Jordan BD, Firer C (2001) Fundamentals of corporate finance. McGraw-Hill Book Company, Sydney

Rudowsky I (2004) Intelligent agents. Commun Assoc Inf Syst 14:275–290

Schach S (2006) Object-oriented and classical software engineering. McGraw-Hill, New York

Shukla R, Shukla M, Misra AK, Marwala T, Clarke WA (2012) Dynamic software maintenance effort estimation modeling using neural network, rule engine, and multi-regression approach. Lect Notes Comput Sci 7336:157–169

Simon H, Rescher N (1966) Cause and counterfactual. Philos Sci 33:323–340

Soares F, Burken J, Marwala T (2006) Neural network applications in advanced aircraft flight control system, a hybrid system, a flight test demonstration. Lect Notes Comput Sci 4234:684–691

Spanos A (2012) Philosophy of econometrics. In: Philosophy of Economics, editor, U. Maki, general editors D. Gabbay, P. Thagard, and J. Woods, Elsevier, Philos Econ 12:329–393

Swann GMP (2008) Putting econometrics in its place: a new direction in applied economics. Edward Elgar Publishing, Brookfield

Tettey T, Marwala T (2006) Controlling interstate conflict using neuro-fuzzy modeling and genetic algorithms. In: Proceedings of the 10th IEEE international conference on intelligent engineering systems, London, 2006, pp 30–44

Tettey T, Marwala T (2006b) Neuro-fuzzy modeling and fuzzy rule extraction applied to conflict management. Lect Notes Comput Sci 4234:1087–1094

Tettey T, Marwala T (2007) Conflict modelling and knowledge extraction using computational intelligence methods. In: Proceedings of the 11th IEEE international conference on intelligent engineering systems, Budapest, Hungary, pp 161–166

Thompson M (1996) Late industrialisers, late democratisers: developmental states in the Asia-Pacific. Third World Q 17:625–647

Tiwari AK (2012) An empirical investigation of causality between producers' price and consumers' price indices in Australia in frequency domain. Econ Model 29:1571–1578

Turova TS (2012) The emergence of connectivity in neuronal networks: from bootstrap percolation to auto-associative memory. Brain Res 1434:277–284

van Aardt B, Marwala T (2005) A study in a hybrid centralised-swarm agent community. In: Proceedings of the IEEE 3rd international conference on computational cybernetics, Mauritius, 2005, pp 169–174

Vapnik V (1998) Statistical learning theory. Wiley-Interscience, New York

Velascoa N, Dejax P, Guéret C, Prins C (2012) A non-dominated sorting genetic algorithm for a bi-objective pick-up and delivery problem. Eng Optim 44:305–325

Wagner T (2002) An agent-oriented approach to industrial automation systems. Lect Notes Comput Sci 2592:314–328

Weisfeld M (2009) The object-oriented thought process. Addison-Wesley, Boston, MA

Woo-Cumings M (1999) The developmental state. Cornell University Press, New York

Wooldridge M (2004) An introduction to multiagent systems. Wiley, New York

Wu Z, Zhang Y, Shen Z, Guan J, Li S, Liu X, Zhang R (2012) System of water supply based on blind source separation. Lect Notes Electr Eng 136:55–59

Xing B, Marwala T (2011) The role of remanufacturing in building a developmental state. Think 33:18–20

Xing B, Gao W-J, Nelwamondo FV, Battle K, Marwala T (2010) Part-machine clustering: the comparison between adaptive resonance theory neural network and ant colony system. Book Ser Lect Notes Electr Eng 67:747–755

Yogo M (2008) Measuring business cycles: a wavelet analysis of economic time series. Econ Lett 100:208–212

Zhang Q, Chen S, Yu C (2012a) Impulsive consensus problem of second-order multi-agent systems with switching topologies. Commun Nonlinear Sci Numer Simul 17:9–16

Zhang S, Wong H-S, Shen Y (2012b) Generalized adjusted rand indices for cluster ensembles. Pattern Recognit 45:2214–2226

Zhou HF, Ni YQ, Ko JM (2011) Structural damage alarming using auto-associative neural network technique: exploration of environment-tolerant capacity and setup of alarming threshold. Mech Syst Signal Process 25:1508–1526

Zhu Q, Yang D, Zhu G (2012) Moving track control system of agriculture oriented mobile robot. Adv Intell Soft Comput 149:179–184

Chapter 2
Techniques for Economic Modeling: Unlocking the Character of Data

Abstract In this chapter techniques for understanding economic data are described. These methods include measures such as the mean, variance, kurtosis, fractals, stationarity, frequency and time-frequency analysis techniques as well as their applications to the understanding economic data.

2.1 Introduction

The widespread availability of economic data has prompted researchers and practitioners to device ways of analyzing these data sets. Some of these data sets are estimated such as the inflation rate or they are priced by the market such as a trading share price. Data analysis is an area that has been around for a long time. Some of the techniques that have been used for estimation, like averages, are now so common that they are used as part of the normal daily conversation lexicology.

The conclusions that are drawn from these data and the statistical parameters that are derived are far reaching. For example, the conclusions that are drawn from the average changes of prices in an economy, also called inflation rate, are used to negotiate salary increases by unionized workers. The implication of a miscalculation of this very vital parameter is far reaching for the workers and the general economy of a country.

Recently, researchers and practitioners have developed decision making tools that are used to assist decision makers make their decisions. For example, a stock trader no longer has to necessarily visually look at complicated charts in order to make his or her decision. Instead, they can use a neural network that is able assist them in making a decision. Of course, this depends on the availability of data and an effective way of handling such data.

Normally before a piece of data is used, it is important to process it. For example, if we were to analyze a time history for the past 50 years of an economy of a particular country, sometimes it is not enough to just calculate the average or variance or any other statistical parameter of this time history. In such a situation,

T. Marwala, *Economic Modeling Using Artificial Intelligence Methods*, Advanced Information and Knowledge Processing, DOI 10.1007/978-1-4471-5010-7_2, © Springer-Verlag London 2013

it might be more advisable to take the time data and convert it into the frequency domain because vital features would be revealed in the frequency domain rather than in the time domain.

This chapter gives a brief account of some of the techniques that are used to analyze data. We consider four domains, where we can analyze economic data and these are time, frequency, time-frequency and fractal domains. In the time domain or the so-called time series framework, we apply statistical concepts of mean, variance and kurtosis to analyze the data. In the frequency domain we use both the Fourier transform to analyze the data; in the time-frequency domain we apply wavelets to analyze the data while in the fractal domain we use the Hurst dimension to analyze the data. Another important factor is the characterization of the stationarity of data. Data is stationary if its characteristics are not changing as a function of time. In this regard, we apply the variance ratio test to characterize the stationarity of the data.

2.2 Time Domain Data

Data can be presented in the time domain which is what makes it known as time series data. In this domain, data is presented as a function of time. For example, we could present the quarterly GDP of the United States from 1 January 1947 to 1 April 2012 as shown in Fig. 2.1. From this figure, it can be observed that the GDP has been consistently growing during the specified time period. From this data, several statistical parameters can be derived and we pay attention to the average, variance and kurtosis.

The data in Fig. 2.1 can be expressed in terms of percentage change of GDP and this is shown in Fig. 2.2.

2.2.1 Average

The average is the measure of the central tendency of data. For N data points, the average $\bar{x}$ of series $x_1, x_2, \ldots, x_N$ can be calculated as follows (Hand 2008):

$$\bar{x} = \frac{1}{N} \sum_i^N x_i \qquad (2.1)$$

Ni et al. (2013) studied the variable length moving average trading rules and its impact on a financial crisis period, while Pavlov and Hurn (2012) tested the profitability of moving-average rules as a portfolio selection strategy and found that, for a wide range of parameters, moving-average rules generate contrarian profits. Chiarella et al. (2012) applied moving averages successfully in a double auction

Fig. 2.1 The quarterly GDP of the United States from 1947 to 2012

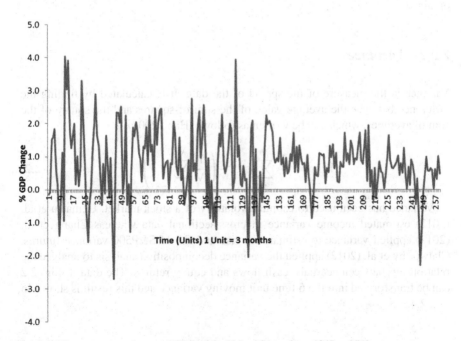

Fig. 2.2 The percentage change of GDP of the United States from 1947 to 2012

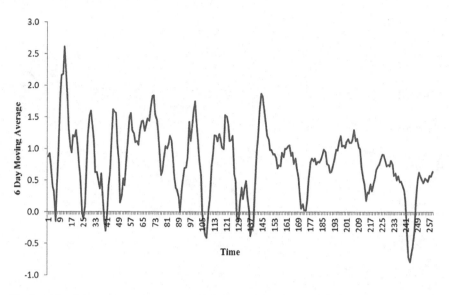

Fig. 2.3 The 6 time unit moving average of the percentage change of GDP of the United States from 1947 to 2012

market. The data in Fig. 2.2 can be further processed by calculating the 6 time unit moving average of the percentage change of GDP and the results in Fig. 2.3 are obtained.

2.2.2 Variance

Variance is the measure of the spread of the data. It is calculated by finding the difference between the average value of the sum-of-squares and the square of the sum of averages, which can be written as follows (Hand 2008):

$$Varx = \frac{1}{N} \sum_{i}^{N} (x_i)^2 - \left(\frac{1}{N} \sum_{i}^{N} (x_i) \right)^2 \tag{2.2}$$

Variance has been used to estimate volatility of a stock market. Uematsu et al. (2012) estimated income variance in cross-sectional data whereas Chang et al. (2012) applied variance to estimate the rise and fall of S&P500 variance futures. Clatworthy et al. (2012) applied the variance decomposition analysis to analyze the relationship between accruals, cash flows and equity returns. The data in Fig. 2.2 can be transformed into the 6 time unit moving variance and this result is shown in Fig. 2.4.

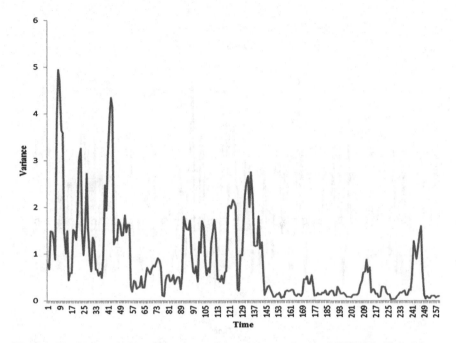

Fig. 2.4 The 6 unit period moving variance of the percentage change of GDP of the United States from 1947 to 2012

From this figure, it can be observed in the earlier years the growth of the GDP was more volatile than in later years.

2.2.3 Kurtosis

There is a need to deal with the occasional spiking of economic data and to achieve this task, Kurtosis is applied. Diavatopoulos et al. (2012) used Kurtosis changes to study the information content prior to earnings announcements for stock and option returns. They observed that changes in Kurtosis predict future stock returns. Some other applications of Kurtosis include in portfolio rankings (di Pierro and Mosevichz 2011), hedging (Angelini and Nicolosi 2010), and risk estimation (Dark 2010). The calculated Kurtosis value is typically normalized by the square of the second moment. A high value of Kurtosis indicates a sharp distribution peak and demonstrates that the signal is impulsive in character. Kurtosis can be written as follows (Hand 2008):

$$K = \frac{1}{N} \sum_{i=1}^{N} \frac{(x_i - \bar{x})^4}{\sigma^4} \tag{2.3}$$

where $\bar{x}$ is the mean and σ is the variance (Fig. 2.5).

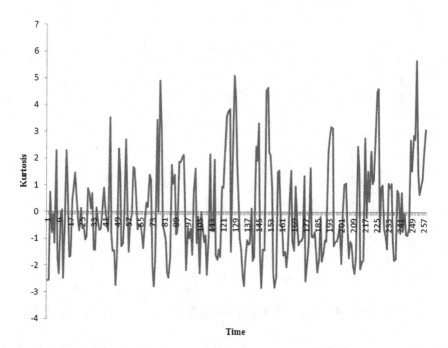

Fig. 2.5 The 6 unit period moving kurtosis of the percentage change of GDP of the United States from 1947 to 2012

2.3 Frequency Doman

The roots of the frequency domain are in the Fourier series, which basically states that every periodic function can be estimated using a Fourier series. This Fourier series is in terms of sine and cosine. This in essence implies that each signal is represented in terms of a series of cycles with different amplitudes and frequencies. Within the economic modeling perspective this implies that we are representing an economic time series as a superposition of cycles. The function $f(x)$ can be estimated using the Fourier series and this is written as follows (Moon and Stirling 1999):

$$f(x) \approx \frac{a_0}{2} + \sum_{n=1}^{N} (a_n \cos(nx) + b_n \sin(nx)), N \geq 0 \qquad (2.4)$$

where

$$a_n = \frac{1}{\pi} \int_{-\pi}^{\pi} f(x) \cos(nx) dx, n \geq 0 \qquad (2.5)$$

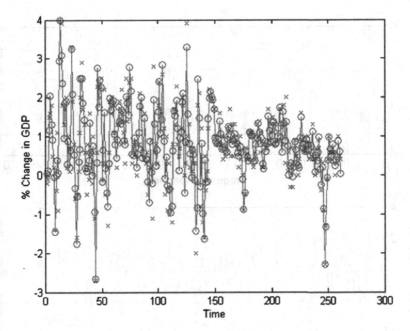

Fig. 2.6 The Fourier series reconstruction of the percentage change of GDP of the United States from 1947 to 2012 (Key: x: 10 terms, o: 100 terms)

and

$$b_n = \frac{1}{\pi} \int_{-\pi}^{\pi} f(x) \sin(nx) dx, n \geq 1 \tag{2.6}$$

The estimation procedure outlined in Eqs. 2.4, 2.5, and 2.6 can be used to estimate Fig. 2.2 and this is shown in Fig. 2.6.

The representation of a signal using sine and cosine functions necessarily implies that a time domain signal can be transformed into the frequency domain. This can be achieved by applying the fast Fourier transform (FFT), which is essentially a computationally efficient technique for calculating the Fourier transform through exploiting the symmetrical relationship of the Fourier transform. If the FFT is applied to the function, $x(t)$, can be written as follows (Moon and Stirling 1999):

$$X(\omega) = \frac{1}{2\pi} \int_{-\infty}^{\infty} x(t) e^{-i\omega t} dt \tag{2.7}$$

where ω is the frequency and t is the time. This relationship can be written as follows in the discrete form:

$$X_k = \sum_{n=0}^{N-1} x_n e^{-i 2\pi k \frac{n}{N}}, k = 0, \ldots, N-1 \tag{2.8}$$

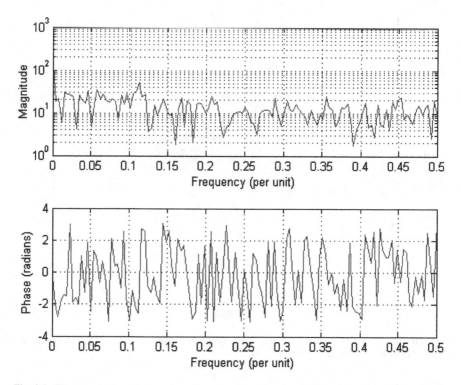

Fig. 2.7 The magnitude and phase from the percentage change of GDP of the United States from 1947 to 2012

The Fourier transform of the signal in Fig. 2.2 is expressed in Fig. 2.7. In this figure, the peaks in the magnitude plot correspond to the cycles and since here we are dealing with an economic data it corresponds business cycle. The phase plot indicates the time lag associated with the magnitude. Grossmann and Orlov (2012) applied frequency domain methods to study exchange rate misalignments. They considered the deviations of Canadian, Japanese, and British spot exchange rates against the US dollar. The results showed that the Plaza Accord and the Euro introduction reduced the volatility of the exchange rate misalignments and extra returns for the Yen and the Pound along nearly all frequency components considered. Tiwari (2012) applied frequency domain methods to study causality between producers' price and consumers' price indices in Australia. The results obtained indicated that consumer price causes producer price at middle frequencies reflecting medium-run cycles, while producer price does not cause consumers price at any frequency. Shahbaz et al. (2012) applied the frequency domain methods to study causality between the Consumer Price Index (CPI) and the Wholesale Price Index (WPI). They observed one directional causal link from the CPI to WPI that varies across frequencies, that is, CPI Granger-causes WPI at all frequencies. Gradojevic (2012) applied frequency domain techniques to analyze foreign exchange order flows. The results indicated that causality depended on the customer type, frequency, and time period.

2.4   Time-Frequency Domain

Time-frequency techniques are methods where it is possible to see what is going on in both the time and frequency domains at the same instance. There is a number of time-frequency techniques and these include Short-time Fourier transform, Wavelet transform, Bilinear time-frequency distribution function (e.g. Wigner distribution function), modified Wigner distribution function, and Gabor–Wigner distribution function (Goupillaud et al. 1984; Delprat et al. 1992; Cohen 1995; Flandrin 1999; Papandreou-Suppappola 2002). In this chapter, we apply wavelet analysis to characterize an economic data in Fig. 2.2. There are many types of wavelets that can be used to analyze a signal and this chapter concentrates on the Morlet type (Chui 1992; Kingsbury 2001).

Zheng and Washington (2012) studied the choice of an optimal wavelet for detecting singularities in traffic and vehicular data. They found that choosing an appropriate wavelet mainly depends on the problem at hand and that the Mexican Hat wavelet offered an acceptable performance in detecting singularities in traffic and vehicular data. Ćmiel (2012) applied the wavelet shrinkage technique to estimate Poisson intensity of the Spektor-Lord-Willis problem. They found that the adaptive estimator gave the optimal rate of convergence over Besov balls to within logarithmic factors. Caraiani (2012) applied wavelets to study the properties of business cycles in Romania between 1991 and 2011 by analyzing the relationship between output and key macroeconomic variables in time and frequency. The results demonstrated that it is possible to separate the influence of definite events. Haven et al. (2012) applied wavelets to de-noise option prices. They demonstrated that the estimation of risk-neutral density functions and out-of-sample price forecasting is significantly improved after noise is removed using the wavelet method. Svensson and Krüger (2012) applied the wavelet method for analysing the mortality and economic fluctuations of Sweden between 1800, whereas Dajcman et al. (2012) successfully applied wavelets to analyze European stock market movement dynamics during financial crises and Hacker et al. (2012) successfully applied wavelets to study the relationship between exchange rates and interest.

In this chapter, we apply the Morlet wavelet which is a wavelet with a complex exponential multiplied by a Gaussian window. It has its roots in the Gabor transform, which applies concepts from quantum physics and Gaussian-windowed sinusoids for time-frequency decomposition and give the best trade-off between spatial and frequency resolution (Connor et al. 2012; Tao and Kwan 2012; Fu et al. 2013; Gu and Tao 2012; Agarwal and Maheshwari 2012). Goupillaud et al. (1984) adapted the Gabor transform to maintain the same wavelet shape over equal octave intervals, resulting in the continuous wavelet transform. The Morlet wavelet has been successfully applied before in a number of areas such as in analyzing earthquake ground motion (Shama 2012), simulating vehicle full-scale crash test (Karimi et al. 2012), seizure detection (Prince and Rani Hemamalini 2012), texture analysis for trabecular bone X-ray images (El Hassani et al. 2012), and fault diagnosis in rolling element bearing (Zhang and Tan 2012). The Morlet wavelet is described mathematically as follows (Goupillaud et al. 1984):

$$\Phi_\sigma(t) = c_\sigma \pi^{-\frac{1}{4}} e^{-\frac{1}{2}t^2} \left(e^{i\sigma t} - \chi_\sigma \right) \tag{2.9}$$

Here $\chi_\sigma = e^{-\frac{1}{2}\sigma^2}$ and is called the admissibility criterion while the normalization constant is:

$$c_\sigma = \left(1 + e^{-\sigma^2} - 2e^{-\frac{3}{4}\sigma^2} \right)^{-\frac{1}{2}} \tag{2.10}$$

The Fourier transform of the Morlet wavelet can be written as follows (Goupillaud et al. 1984):

$$\Phi_\sigma(\omega) = c_\sigma \pi^{-\frac{1}{4}} \left(e^{-\frac{1}{2}(\sigma-\omega)^2} - \chi_\sigma e^{-\frac{1}{2}\omega^2} \right) \tag{2.11}$$

The variable σ permits the trade-off between time and frequency resolutions. The results obtained when the Morlet wavelets are used to analyze Fig. 2.2 are shown in Fig. 2.8.

2.5 Fractals

Fractal analysis is a method of defining complex shapes and numerous techniques for estimating fractal dimensions have been proposed (Marwala 2012). As described by Lunga (2007) and Lunga and Marwala (2006), fractal dimensions of an object are an indicator of the degree to which the object occupies space. Alternatively, a fractal dimension of a time series expresses how turbulent the time series is and also quantifies the extent to which the time series is scale-invariant (Lunga and Marwala 2006; Lunga 2007). The technique applied to approximate fractal dimensions using the Hurst exponent for a time series is known as the rescaled range (R/S) analysis and was proposed by Hurst (1951).

Lunga and Marwala (2006) successfully applied time series analysis applying fractal theory and online ensemble classifiers to model the stock market, while Nelwamondo et al. (2006a) applied fractals successfully for early classifications of bearing faults. Nelwamondo et al. (2006b) applied a multi-scale fractal dimension for speaker identification systems, while Nelwamondo et al. (2006c) applied fractals for improving speaker identification rates.

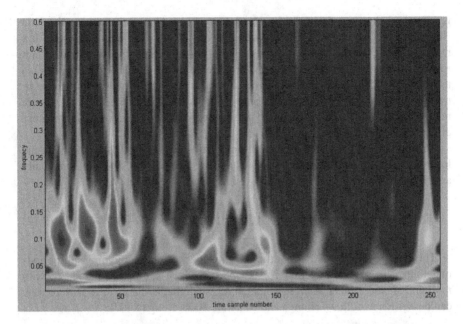

Fig. 2.8 The wavelet transform of the percentage change of GDP of the United States from 1947 to 2012

Kristoufek (2012) studied the influence of the fractal markets hypothesis for liquidity and investment horizons on predictions of the dynamics of the financial markets during turbulences such as the Global Financial Crisis of late 2000s. They observed that fractal markets hypothesis predicted the observed characteristics sufficiently.

Krištoufek and Vošvrda (2012) used the fractal dimension, Hurst exponent, and entropy to quantify capital markets efficiency. The results obtained indicated that the efficient market was dominated by local inefficiencies and stock indices of the most developed countries were the most efficient capital markets.

Shin et al. (2012) applied fractals to organize distributed and decentralized manufacturing resources. The proposed technique was observed to reduce problem complexity through iterative decomposition of problems in resource management.

2.5.1 The Rescaled Range (R/S) Methodology

As described by Lunga and Marwala (2006) as well as Lunga (2007), this section describes a method for approximating the quality of a time series signal to identify the intervals that are vital to classify a time signal. The rescaled range (R/S) analysis is a method that was proposed by Hurst (1951) to control reservoir on the Nile River dam project in 1907. The objective was to identify the optimal design of a reservoir from data of measured river discharges. A desirable reservoir does not run dry or overflow. Hurst proposed a statistical quantity, the Hurst exponent (H), a technique that can be applied to categorize time series signals into random and non-random series. By applying the R/S analysis, it is possible to identify the average non-periodic cycle and the measure of persistence in trends because of long memory effects (Skjeltorp 2000).

To implement R/S when we have a time series of length M is to calculate the logarithm of a ratio with length $N = M - 1$ and this can be expressed mathematically as follows (Lunga 2007):

$$N_i = \log \left(\frac{M_{i+1}}{M_i} \right), \ i = 1, 2, 3, \ldots, (M-1) \tag{2.12}$$

The average is estimated by splitting the time period into T adjoining sub-periods of length j, in such a way that $T * j = N$, with each sub-period named I_t, with $t = 1, 2 \ldots T$ and each element in I_t named $N_{k,t}$ such that $k = 1, 2, \ldots j$. This average can be written as follows (Lunga and Marwala 2006; Lunga 2007):

$$e_t = \frac{1}{j} \sum_{k=1}^{j} N_{k,t} \tag{2.13}$$

Therefore, e_t is the average value of the N_i enclosed in sub-period I_t of length j. The time series of accrued data $X_{k,t}$ can be calculated from the mean for each sub-period I_t, as follows (Lunga 2007):

$$X_{t,t} = \sum_{i=1}^{k} (N_{k,t} - e_t) k = 1, 2, 3, \ldots, j \tag{2.14}$$

The range of the time series in relation to the mean within each sub-period can be mathematically written as follows (Lunga 2007):

$$R_{T_i} = \max (X_{k,t}) - \min (X_{k,t}), 1 < k < j \tag{2.15}$$

The standard deviation of each sub-period can be calculated as follows (Lunga 2007):

$$S_{I_t} = \sqrt{\frac{1}{j} \sum_{i=1}^{k} (N_{i,t} - e_t)^2} \tag{2.16}$$

The range of each sub-period R_{Tt} can be rescaled by the respective standard deviation S_{It} and is done for all the T sub-intervals in the series and the following average R/S value is obtained (Lunga 2007):

$$e_t = \frac{1}{T} \sum_{t=1}^{T} \left(\frac{R_{It}}{S_{It}} \right) \tag{2.17}$$

The computation from the Eqs. 2.12, 2.13, 2.14, 2.15, 2.16 and 2.17 can be recurred for different time ranges and this can be realized by sequentially increasing j and reiterating the computation until all j values are covered. After computing the R/S values for a large range of different time ranges j, $log(R/S)_j$ can be plotted versus $log(n)$. The Hurst exponent (H) can be estimated by conducting a least squares linear regression with $log(R/S)$ as the dependent variable and $log(n)$ as the independent variable and H is the slope of the regression (Maragos and Potamianos 1999; Hurst 1951; Hurst et al. 1965). The fractal dimension and the Hurst exponent are related as follows (Lunga 2007; Wang et al. 2000):

$$D_f = 2 - H \tag{2.18}$$

There are other techniques for estimating the fractal dimensions and these include using the Box counting dimension (Falconer 1952; Nelwamondo et al. 2006a; Marwala 2012), the Hausdorff dimension (Falconer 1952) and Minkowski-Boulig and dimension (Schroeder 1991).

2.5.2 The Hurst Interpretation

In fractal theory, when $H \in (0.5; 1]$ then the time series is persistent and is described by long memory effects on all time scales (Lo 1991; Gammel 1998). Within the context of the stock market, this means that all prices are correlated with all future hourly price changes (Beran 1994). Persistence indicates that if the time series has been up or down in the previous time, then there is a probability that it will remain up and down, respectively, in the future time. The advantage of the tendency to reinforce behavior, or persistence, increases as H tends to 1. The influence of the present on the future can be articulated as a correlation function G as follows (Lunga 2007):

$$G = 2^{2H-1} - 1 \tag{2.19}$$

When $H = 0.5$ then $G = 0$ and the time series is uncorrelated. Nevertheless, when $H = 1$ then $G = 1$, representing ideal positive correlation. Alternatively, when $H \in [0; 0.5)$ then the time series signal is anti-persistent and this implies that every

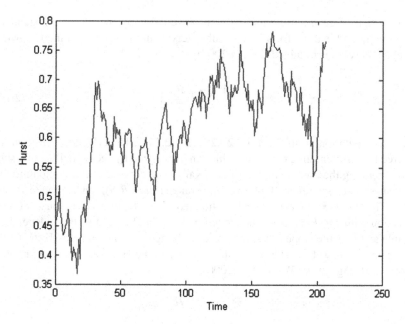

Fig. 2.9 Hurst of the GDP of the United States from 1947 to 2012

time a time series has been up in the last period, it is more possibly going to be down in the next period. Therefore, an anti-persistent time series is spiky than a pure random walk.

The Hurst factor was used to analyse the data in Fig. 2.1 and the results in Fig. 2.9 were obtained. These results were obtained by calculating the H over a period of 50 time units. Here, a unit corresponds to 3 months. The results in Fig. 2.9 indicate that the data during this period was largely persistent, which indicates that the signal has long memory effects. The results in Fig. 2.10 indicate that this data was anti-persistent and this means that every time the GDP change has been up in the previous period it is most likely going to be down in the next period.

2.6   Stationarity

A stationary process is a stochastic process whose joint probability distribution does not vary when shifted in space or time. Therefore, if certain parameters (for instance, the mean and variance) can be estimated, then they do not change over space or time (Priestley 1988). By the same logic, a non-stationary process is a process whose joint probability distribution varies when shifted in space or time. There are many techniques that have been proposed to evaluate whether a given signal is stationary or not.

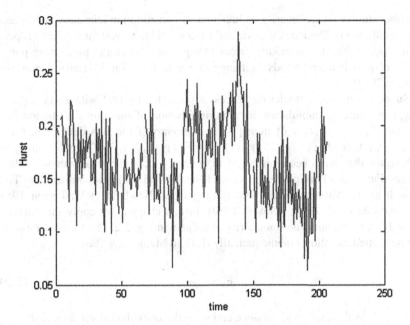

Fig. 2.10 Hurst of the percentage change of the GDP of the United States from 1947 to 2012

Kiremire and Marwala (2008) proposed the stationarity index, which is a quantification of similarities of the auto correlation integral of a subdivision of a time series and the cross-correlation of that subdivision with others of the same time series. This quantification of similarity is a measure of the stationarity of the time series and, consequently, can be applied to detect and quantify non-stationarity.

The index was successfully applied in the analysis of eleetrocardiogram (ECG) and electroencephalogram (EEG) signals to detect the variations in the dynamics of the signals and the incidence of several events. The index showed sensitivity to variations in the dynamics shown in ECG signals that were the result of partial epileptic seizures. Zhou and Kutan (2011) applied non-linear unit root tests and recursive analysis to evaluate the relationship between the stationarity of real exchange rates and different currencies, different sample periods, and different countries. The results showed that a stationary real exchange rate is sensitive to sample period but not the currencies.

Caporale and Paxton (2013) investigated inflation stationarity in Brazil, Argentina, Chile, Mexico, and Bolivia from 1980 to 2004. They tested for structural interruptions in inflation, explained the interruptions using changes in monetary policies, and tested the relationship between structural interruptions and non-stationarity results.

Zhou (2013) successfully studied the nonlinearity and stationarity of inflation rates in the Euro-zone countries, whereas Fallahi (2012) successfully studied the stationarity of consumption-income ratios (CIR) and observed that the CSI was non-stationary in most of the countries.

Other studies on stationarity include energy consumption (Hasanov and Telatar 2011), inflation in Mexico (Caporale and Paxton 2011), interest rate in the European Union (Zhou 2011), commodity prices (Yang et al. 2012) and purchasing power parity of post-Bretton Woods exchange rate data for 20 industrialized countries (Amara 2011).

Suppose we have a stochastic process represented by $\{Y_t\}$ with $F_Y(y_{t_1+\zeta}, \ldots, y_{t_n+\zeta})$ indicating a cumulative distribution function of the joint distribution $\{Y_t\}$ at times $\{t_1 + \zeta, \ldots, t_n + \zeta\}$ then $\{Y_t\}$ is stationary if for all values of n, ζ and $\{t_1, \ldots, t_n\}$ then $F_Y(y_{t_1+\zeta}, \ldots, y_{t_n+\zeta}) = F_Y(y_{t_1}, \ldots, y_{t_n})$. There are many of techniques that have been proposed to quantify stationarity and these include Dickey-Fuller and Phillips-Perron Tests, Kwiatkowski-Phillips-Schmidt-Shin Test, as well as the Variance Ratio Test (Granger and Newbold 1974; Perron 1988; Kwiatkowski et al. 1992; Schwert 1989). In this chapter, we apply the Variance Ratio Test to determine the stationarity of a signal in Fig. 2.2. The variance ratio (F) can be written as follows mathematically (Lo and MacKinlay 1989):

$$F = \frac{Ve}{Vu} \tag{2.20}$$

Here, E_v is the explained variance and U_v is the unexplained variance and:

$$V_e = \frac{\sum_i n_i \left(\bar{X}_i - \bar{X} \right)^2}{K - 1} \tag{2.21}$$

and

$$V_u = \frac{\sum_{ij} \left(X_{ij} - \bar{X}_i \right)^2}{N - K} \tag{2.22}$$

Here, $\bar{X}_i$ indicates the sample mean of the ith group, n_i is the number of observations in the ith group, a $\bar{X}$ indicates the complete mean of the data, X_{ij} is the jth observation in the ith out of K groups and N is the overall sample size. If the variability ratio is 1, then the data is following a random walk, if it is larger than 1, then it is trending and, therefore, non-stationary and if it is less than one, then it shows a mean reversal meaning than changes in one direction leads to probably changes in the opposite direction. The results obtained when we tested for stationarity using the variance ratio test are shown in Fig. 2.11 for the data in Figs. 2.1 and 2.12 for the data in Fig. 2.2.

These results were obtained by analyzing a moving window of 50 units with each unit corresponding to 3 months. These results indicate that the raw values of the GDP in Fig. 2.1 is non-stationary while the percentage change of GDP in Fig. 2.2 is stationary.

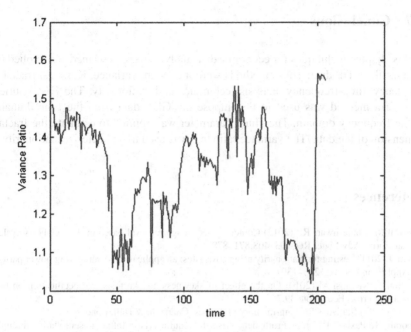

Fig. 2.11 Variance ratio of the GDP of the United States from 1947 to 2012

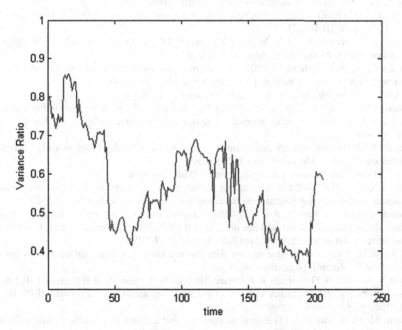

Fig. 2.12 Variance ratio of the percentage change of the GDP of the United States from 1947 to 2012

2.7 Conclusions

In this chapter, techniques for economic data analysis were described and applied to analyze the GDP data. These methods were the mean, variance, Kurtosis, fractals, frequency, time-frequency analysis techniques and stationarity. The Fast Fourier transform method was used to decompose the GDP data from the time domain to the frequency domain. The Hurst parameter was applied to estimate the fractal dimension of the data. The variance ratio test was used to characterize stationarity.

References

Agarwal M, Maheshwari RP (2012) Content based image retrieval based on Log Gabor wavelet transform. Adv Mater Res 403–408:871–878

Amara J (2011) Testing for stationarity using covariates: an application to purchasing power parity. Appl Econ Lett 18:1295–1301

Angelini F, Nicolosi M (2010) On the effect of skewness and kurtosis misspecification on the hedging error. Econ Note 39:203–226

Beran J (1994) Statistics for long-memory processes. Chapman & Hall, London

Caporale T, Paxton J (2011) From debt crisis to tequila crisis: inflation stationarity through Mexico's turbulent decades. Appl Econ Lett 18:1609–1612

Caporale T, Paxton J (2013) Inflation stationarity during Latin American inflation: insights from unit root and structural break analysis. Appl Econ 45:2001–2010

Caraiani P (2012) Stylized facts of business cycles in a transition economy in time and frequency. Econ Model 29:2163–2173

Chang C-L, Jimenez-Martin J-A, McAleer M, Amaral TP (2012) The rise and fall of S&P500 variance futures. N Am J Econ Finance (Article in Press)

Chiarella C, He X-Z, Pellizzari P (2012) A dynamic analysis of the microstructure of moving average rules in a double auction market. Macroecon Dyn 16:556–575

Chui CK (1992) An introduction to wavelets. Academic, San Diego

Clatworthy MA, Pong CKM, Wong WK (2012) Auditor quality effects on the relationship between accruals, cash flows and equity returns: a variance decomposition analysis. Account Bus Res 42:419–439

Ćmiel B (2012) Poisson intensity estimation for the Spektor-Lord-Willis problem using a wavelet shrinkage approach. J Multivar Anal 112:194–206

Cohen L (1995) Time-frequency analysis. Prentice-Hall, New York

Connor EF, Li S, Li S (2012) Automating identification of avian vocalizations using time-frequency information extracted from the Gabor transform. J Acoust Soc Am 132:507–517

Dajcman S, Festic M, Kavkler A (2012) European stock market comovement dynamics during some major financial market turmoils in the period 1997–2010 – a comparative DCC-GARCH and wavelet correlation analysis. Appl Econ Lett 19:1249–1256

Dark JG (2010) Estimation of time varying skewness and kurtosis with an application to value at risk. Stud Nonlinear Dyn Econom 14:art. no. 3

Delprat N, Escudié B, Guillemain P, Kronland-Martinet R, Tchamitchian P, Torrésani B (1992) Asymptotic wavelet and Gabor analysis: extraction of instantaneous frequencies. IEEE Trans Inf Theory 38:644–664

di Pierro M, Mosevichz J (2011) Effects of skewness and kurtosis on portfolio rankings. Quant Finance 11:1449–1453

Diavatopoulos D, Doran JS, Fodor A, Peterson DR (2012) The information content of implied skewness and kurtosis changes prior to earnings announcements for stock and option returns. J Bank Finance 36:786–802

El Hassani ASEB, El Hassouni M, Jennane R, Rziza M, Lespessailles E (2012) Texture analysis for trabecular bone X-ray images using anisotropic morlet wavelet and Rényi entropy. Lect Note Comput Sci 7340:290–297

Falconer K (1952) Fractal geometry: mathematical foundations and application. Wiley, New York

Fallahi F (2012) The stationarity of consumption-income ratios: evidence from bootstrapping confidence intervals. Econ Lett 115:137–140

Flandrin P (1999) Time-frequency/time-scale snalysis, vol 10, Wavelet analysis and its applications. Elsevier, Amsterdam

Fu YB, Chui CK, Teo CL (2013) Accurate two-dimensional cardiac strain calculation using adaptive windowed fourier transform and gabor wavelet transform. Int J Comput Assist Radiol Surg 8(1):135–144

Gammel B (1998) Hurst's rescaled range statistical analysis for pseudorandom number generators used in physical simulations. Phys Rev E 58:2586–2597

Goupillaud P, Grossman A, Morlet J (1984) Cycle-octave and related transforms in seismic signal analysis. Geoexploration 23:85–102

Gradojevic N (2012) Frequency domain analysis of foreign exchange order flows. Econ Lett 115:73–76

Granger CWJ, Newbold P (1974) Spurious regressions in econometrics. J Econom 2:111–120

Grossmann A, Orlov AG (2012) Exchange rate misalignments in frequency domain. Int Rev Econ Finance 24:185–199

Gu JJ, Tao L (2012) Filterbank and DFT based fast parallel discrete Gabor transform for image representation. Adv Mater Res 461:444–447

Hacker SR, Karlsson HK, Månsson K (2012) The relationship between exchange rates and interest rate differentials: a wavelet approach. World Econ 35:1162–1185

Hand DJ (2008) Statistics: a very short introduction. Oxford University Press, Oxford

Hasanov M, Telatar E (2011) A re-examination of stationarity of energy consumption: evidence from new unit root tests. Energy Policy 39:7726–7738

Haven E, Liu X, Shen L (2012) De-noising option prices with the wavelet method. Eur J Oper Res 222:104–112

Hurst HE (1951) Long term storage capacity of reservoirs. Trans Am Soc Eng 116:770–799

Hurst HE, Black RP, Simaika YM (1965) Long-term storage: an experimental study. Constable, London

Karimi HR, Pawlus W, Robbersmyr KG (2012) Signal reconstruction, modeling and simulation of a vehicle full-scale crash test based on morlet wavelets. Neurocomputing 93:88–99

Kingsbury NG (2001) Complex wavelets for shift invariant analysis and filtering of signals. J Appl Comput Harmon Anal 10:234–253

Kiremire BBE, Marwala T (2008) Non-stationarity detection: a stationarity index approach. In: Proceedings of the IEEE international congress on image and signal processing, Sanya, 2008, pp 373–378

Kristoufek L (2012) Fractal markets hypothesis and the global financial crisis: scaling, investment horizons and liquidity, Adv Complex Syst 15:art. no. 1250065

Krištoufek L, Vošvrda M (2012) Capital markets efficiency: fractal dimension, hurst exponent and entropy. Politicka Ekonomie 60:208–221

Kwiatkowski D, Phillips PCB, Schmidt P, Shin Y (1992) Testing the null hypothesis of stationarity against the alternative of a unit root. J Econom 54:159–178

Lo A (1991) Long-term memory in stock market prices. Econometrica 59:1279–1313

Lo AW, MacKinlay AC (1989) The size and power of the variance ratio test. J Econom 40:203–238

Lunga D (2007) Time series analysis using fractal theory and ensemble classifiers with application to atock portfolio optimization. Masters dissertation, University of the Witwatersrand, Johannesburg

Lunga D, Marwala T (2006) Time series analysis using fractal theory and online ensemble classifiers. Lect Note Artif Intell 4304:312–321

Maragos P, Potamianos A (1999) Fractal dimensions of speech sounds: computation and application to automatic speech recognition. J Acoust Soc Am 105(3):1925–1932

Marwala T (2012) Condition monitoring using computational intelligence methods. Springer, London

Moon TK, Stirling WC (1999) Mathematical methods and algorithms for signal processing. Prentice Hall, New York

Nelwamondo FV, Mahola U, Marwala T (2006a) Multi-scale fractal dimension for speaker identification system. Trans Syst 5:1152–1157

Nelwamondo FV, Marwala T, Mahola U (2006b) Early classifications of bearing faults using hidden Markov models, Gaussian mixture models, Mel-frequency cepstral coefficients and fractals. Int J Innov Comput Inf Control 2:1281–1299

Nelwamondo FV, Mahola U, Marwala T (2006c) Improving speaker identification rate using fractals. In: Proceedings of the IEEE international joint conference on neural networks, Vancouver, 2006, pp 5870–5875

Ni Y-S, Lee J-T, Liao Y-C (2013) Do variable length moving average trading rules matter during a financial crisis period? Appl Econ Lett 20:135–141

Papandreou-Suppappola A (2002) Applications in time-frequency signal processing. CRC Press, Boca Raton

Pavlov V, Hurn S (2012) Testing the profitability of moving-average rules as a portfolio selection strategy. Pac Basin Finance J 20:825–842

Perron P (1988) Trends and random walks in macroeconomic time series: further evidence from a new approach. J Econ Dyn Control 12:297–332

Priestley MB (1988) Non-linear and non-stationary time series analysis. Academic, Waltham

Prince PGK, Rani Hemamalini R (2012) Seizure detection using parameter estimation and Morlet wavelet transform. Commun Comput Inf Sci 270:674–679

Schroeder M (1991) Fractals, chaos, power laws: minutes from an infinite paradise. W. H. Freeman, New York

Schwert W (1989) Tests for unit roots: a Monte Carlo investigation. J Bus Econ Stat 7:147–159

Shahbaz M, Tiwari AK, Tahir MI (2012) Does CPI Granger-cause WPI? New extensions from frequency domain approach in Pakistan. Econ Model 29:1592–1597

Shama A (2012) Spectrum compatible earthquake ground motions by morlet wavelet. In: Proceedings of the 20th analysis and computation specialty conference, Chicago, 2012, pp 163–172

Shin M, Ryu K, Ryu T, Bae S (2012) A management framework for autonomous and intelligent resources in distributed manufacturing systems: a fractal-based approach. Adv Sci Lett 14:203–208

Skjeltorp J (2000) Scaling in the Norwegian stock market. Physica A 283:486–528

Svensson M, Krüger NA (2012) Mortality and economic fluctuations: evidence from wavelet analysis for Sweden 1800–2000. J Popul Econ 25:1215–1235

Tao L, Kwan HK (2012) Multirate-based fast parallel algorithms for 2-D DHT-based real-valued discrete Gabor transform. IEEE Trans Image Process 21:3306–3311

Tiwari AK (2012) An empirical investigation of causality between producers' price and consumers' price indices in Australia in frequency domain. Econ Model 29:1571–1578

Uematsu H, Mishra AK, Powell RR (2012) An alternative method to estimate income variance in cross-sectional data. Appl Econ Lett 19:1431–1436

Wang F, Zheng F, Wu W (2000) A C/V segmentation for Mandarin speech based on multi-scale fractal dimension. Int Conf Spok Lang Process 4:648–651

Yang C-H, Lin C-T, Kao Y-S (2012) Exploring stationarity and structural breaks in commodity prices by the panel data model. Appl Econ Lett 19:353–361

Zhang R-G, Tan Y-H (2012) Fault diagnosis of rolling element bearings based on optimal Morlet wavelet and hidden Markov model. J Vib Shock 31:5–8, +27

Zheng Z, Washington S (2012) On selecting an optimal wavelet for detecting singularities in traffic and vehicular data. Trans Res Part C Emerg Technol 25:18–33

Zhou S (2011) Nonlinear stationarity of real interest rates in the EMU countries. J Econ Stud 38:691–702

Zhou S (2013) Nonlinearity and stationarity of inflation rates: evidence from the euro-zone countries. Appl Econ 45:849–856

Zhou S, Kutan AM (2011) Is the evidence for PPP reliable? A sustainability examination of the stationarity of real exchange rates. J Bank Finance 35:2479–2490

Chapter 3
Automatic Relevance Determination in Economic Modeling

Abstract This chapter introduces the Bayesian and the evidence frameworks to construct an automatic relevance determination method. These techniques are described in detail, relevant literature reviews are conducted, and their use is justified. The automatic relevance determination technique is then applied to determine the relevance of economic variables that are essential for driving the consumer price index. Conclusions are drawn and are explained within the context of economic sciences.

3.1 Introduction

This chapter presents the Bayesian and the evidence frameworks to create an automatic relevance determination technique. These methods are explained in detail, and pertinent literature reviews are conducted. The automatic relevance determination method is then applied to establish the relevance of economic variables that are critical for driving the consumer price index (CPI).

Shutin et al. (2012) introduced incremental reformulated automatic relevance determination and related this to the incremental version of sparse Bayesian learning. The fast marginal likelihood maximization procedure is an incremental method where the objective function is optimized with respect to the parameters of a single component given that the other parameters are fixed. The procedure is then demonstrated to relate to a series of re-weighted convex optimization problems.

Huang et al. (2012) introduced stochastic optimization using an automatic relevance determination (ARD) prior model and applied this for Bayesian compressive sensing. Compressive sensing is a unique data acquisition procedure where the compression is conducted during the sampling process. In condition monitoring systems, original data compression methods such as compressive sensing are required to decrease the cost of signal transfer and storage. Huang et al. (2012) introduced

T. Marwala, *Economic Modeling Using Artificial Intelligence Methods*, Advanced Information and Knowledge Processing, DOI 10.1007/978-1-4471-5010-7_3,
© Springer-Verlag London 2013

Bayesian compressive sensing (BCS) for condition monitoring of signals. The results obtained from the improved BCS technique were better than conventional BCS reconstruction procedures.

Böck et al. (2012) used the ARD method for a hub-cantered gene network reconstruction procedure and exploited topology of the of gene regulatory networks by using a Bayesian network. The proposed technique was applied to a large publicly available dataset was able to identify several main hub genes.

Jacobs (2012) applied Bayesian support vector regression with an ARD kernel for modeling antenna input characteristics. The results indicated that Bayesian support vector regression was appropriate for highly non-linear modeling tasks. They observed that the Bayesian framework allowed efficient training of the multiple kernel ARD hyper-parameters.

Shutin et al. (2011) proposed fast variational sparse Bayesian learning with ARD for superimposed signals. They showed that a fast version of variational sparse Bayesian learning can be built using stationary points of the variational update factors with non-informative ARD hyper-priors.

Zhang et al. (2010) applied the Gaussian process classification using ARD for synthetic aperture radar target recognition. The method they proposed implemented kernel principal component analysis to identify sample features and applied target recognition using Gaussian process classification with an ARD function. When they compared this technique to the k-Nearest Neighbor clustering method, a Naïve Bayes classifier, and a Support Vector Machine, the proposed technique was found to be able to automatically select an appropriate model and to optimize hyper-parameters.

Lisboa et al. (2009) applied a neural network model and Bayesian regularization with the typical approximation of the evidence to create an automatic relevance determination system. The model was applied to local and distal recurrence of breast cancer.

Mørupa and Hansena (2009) applied automatic relevance determination for multi-way models, while Browne et al. (2008) applied ARD for identifying thalamic regions concerned in schizophrenia. Other successful applications of ARD include in earthquake early warning systems (Oh et al. 2008), feature correlations investigation (Fu and Browne 2008), ranking the variables to determine ischaemic episodes (Smyrnakis and Evans 2007), the estimation of relevant variables in multichannel EEG (Wu et al. 2010) and the estimation of relevant variables in classifying ovarian tumors (Van Calster et al. 2006).

3.2 Mathematical Framework

The automatic relevance determination technique is a process applied to evaluate the relevance of each input variable in its ability to predict a particular phenomenon. In this chapter, we apply ARD to determine the relevance of economic variables in predicting the CPI. ARD achieves this task of ranking input variables by optimizing the hyper-parameters to maximize the evidence in the Bayesian framework (Marwala and Lagazio 2011). As already described by Marwala and Lagazio (2011), Wang and Lu (2006) successfully applied the ARD technique to approximate influential variables in modeling the ozone layer, while Nummenmaa et al. (2007) successfully applied ARD to a dataset where a male subject was presented with uncontaminated tone and checker board reversal stimuli, individually and in combination, in employing a magnetic resonance imaging-based cortical surface model. Ulusoy and Bishop (2006) applied ARD to categorize relevant features for object recognition of 2D images. One part of the ARD as applied in this chapter is neural networks, the topic of the next sub-section. It should be borne in mind that the discourse of ARD is not only limited to neural networks and can, within the context of artificial intelligence, include subjects such as support vector machines, Gaussian mixture models, and many other models.

3.2.1 Neural Networks

This section gives a summary of neural networks in the context of economic modeling (Leke and Marwala 2005; Lunga and Marwala 2006). A neural network is an information processing procedure that is inspired by the way that biological nervous systems, like the human brain, process information (Marwala and Lagazio 2011). It is a computer-based mechanism that is aimed at modeling the way in which the brain processes a specific function of consideration (Haykin 1999; Marwala and Lagazio 2011).

The neural network technique is a powerful tool that has been successfully used in mechanical engineering (Marwala and Hunt 1999; Vilakazi and Marwala 2007; Marwala 2012), civil engineering (Marwala 2000, 2001), aerospace engineering (Marwala 2003), biomedical engineering (Mohamed et al. 2006; Russell et al. 2008, 2009a, b), finance (Leke and Marwala 2005; Patel and Marwala 2006; Hurwitz and Marwala 2011; Khoza and Marwala 2012), statistics (Marwala 2009), intrusion detection (Alsharafat 2013), telecommunications (Taspinar and Isik 2013) and political science (Marwala and Lagazio 2011).

Meena et al. (2013) applied fuzzy logic and neural networks for gender classification in speech recognition. To train fuzzy logic and neural networks, features that included the pitch of the speech were used and the tests showed good results. Nasir et al. (2013) applied a multilayer perceptron and simplified fuzzy ARTMAP neural networks for the classification of acute leukaemia cells. The cells were classified into lymphoblast, myoblast, and normal cell; to categorize the severity of leukaemia types, and the results obtained gave good classification performance. Shaltaf and Mohammad (2013) applied a hybrid neural network and maximum likelihood based estimation of chirp signal parameters and observed that a hybrid of neural networks and maximum likelihood gradient based optimization gave accurate parameter approximation for large signal to noise ratio.

In this chapter, neural networks are viewed as generalized regression models that can model any data, linear or non-linear. As described by Marwala and Lagazio (2011), a neural network is made up of four main constituents (Haykin 1999; Marwala 2012):

- the processing units u_j, where each u_j has a particular activation level $a_j(t)$ at any given time;
- weighted inter-connections between several processing units. These inter-connections control how the activation of one unit influences the input for another unit;
- an activation rule, which takes input signals at a unit to yield a new output signal; and
- a learning rule that stipulates how to regulate the weights for a given input/output pair (Haykin 1999).

Neural networks are able to derive meaning from complex data and, therefore, can be used to extract patterns and detect trends that are too complex to be detected by many other computer approaches (Hassoun 1995; Marwala 2012). A trained neural network can be viewed as an expert in the type of information it has been set to discover (Yoon and Peterson 1990; Valdés et al. 2012; Sinha et al. 2013). A trained neural network can be used to predict given new circumstances. Neural networks have been applied to model a number of non-linear applications because of their capacity to adapt to non-linear data (Leke et al. 2007; Martínez-Rego et al. 2012).

The topology of neural processing units and their inter-connections can have a deep influence on the processing capabilities of neural networks. Accordingly, there are many different connections that describe how data flows between the input, hidden, and output layers.

There are different kinds of neural network topologies and these include the multi-layer perceptron (MLP) and the radial basis function (RBF) (Bishop 1995; Sanz et al. 2012; Prakash et al. 2012). In this chapter, the MLP was applied to identify the relationship between economic variables and CPI. The motivation for using the MLP was because it offers a distributed representation with respect to the input space due to cross-coupling between hidden units, while the RBF gives only local representation (Bishop 1995; Marwala and Lagazio 2011).

Ikuta et al. (2012) applied the MLP for solving two-spiral problem whereas Rezaeian-Zadeh et al. (2012) applied the MLP and RBF for hourly temperature prediction. Wu (2012) applied a multi-layer perceptron neural network for scattered point data surface reconstruction while Li and Li (2012) implemented the MLP in a hardware application.

The conventional MLP contains hidden units and output units and normally has one hidden layer. The bias parameters in the first layer are presented as mapping weights from an extra input having a fixed value of $x_0 = 1$. The bias parameters in the second layer are displayed as weights from an extra hidden unit, with the activation fixed at $z_0 = 1$. The model in Fig. 3.1 can take into account the intrinsic dimensionality of the data. Models of this form can approximate any continuous function to arbitrary accuracy if the number of hidden units M is adequately large.

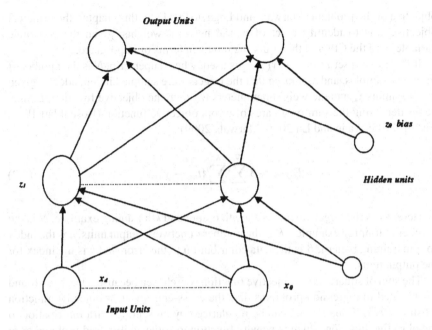

Fig. 3.1 Feed-forward multi-layer perceptron network having two layers of adaptive weights (Reprinted with permission from Marwala 2009; Marwala and Lagazio 2011)

The size of the MLP may be increased by permitting for a number of layers, however, it has been verified by the universal approximation theorem (Cybenko 1989) that a two-layered design is suitable for the MLP. Because of this theorem, in this chapter, the two-layered network shown in Fig. 3.1 was chosen. The relationship between CPI, y, and economic variables, x, may be written as follows (Bishop 1995; Marwala 2012):

$$y_k = f_{outer}\left(\sum_{j=1}^{M} w_{kj}^{(2)} f_{inner}\left(\sum_{i=1}^{d} w_{ji}^{(1)} x_i + w_{j0}^{(1)}\right) + w_{k0}^{(2)}\right) \qquad (3.1)$$

Here, $w_{ji}^{(1)}$ and $w_{ji}^{(2)}$ specify neural network weights in the first and second layers, respectively, going from input i to hidden unit j, M is the number of hidden units, d is the number of output units while $w_{j0}^{(1)}$ specifies the bias for the hidden unit j. This chapter uses a hyperbolic tangent function for the function $f_{inner}(\bullet)$. The function $f_{outer}(\bullet)$ is linear because the problem we are handling is a regression problem (Bishop 1995; Marwala and Lagazio 2011).

On training regular neural networks, the network weights are identified while the training of probabilistic neural networks identifies the probability distributions of the network weights. An objective function must be selected to identify the weights in Eq. 3.1. An *objective function* is a mathematical representation of the global

objective of the problem (Marwala and Lagazio 2011). In this chapter, the principal
objective was to identify a set of neural network weights, given the economic
variables and the CPI and then rank the inputs in terms of importance.

If the training set $D = \{x_k, y_k\}_{k=1}^{N}$ is used, where superscript N is the number of
training examples, and assuming that the targets y are sampled independently given
the kth inputs x_k and the weight parameters w_{kj} then the objective function, E, may
be written using the sum of squares of errors objective function (Rosenblatt 1961;
Bishop 1995; Wang and Lu 2006; Marwala 2009):

$$E_D = -\beta \sum_{n=1}^{N} \sum_{k=1}^{K} \{t_{nk} - y_{nk}\}^2 \qquad (3.2)$$

Here, t_{nk} is the target vector for the nth output and kth training example, N is the
number of training examples, K is the number of network output units, n is the index
for the training pattern, β is the data contribution to the error, and k is the index for
the output units.

The sum of squares error objective function was chosen because it has been found
to be suited to regression problems than the cross-entropy error objective function
(Bishop 1995). Equation 3.2 can be regularized by presenting extra information to
the objective function which is penalty function to solve an ill-posed problem or to
prevent over-fitting by safeguarding smoothness of the solution to reach a trade-off
between complexity and accuracy using (Bishop 1995; Marwala and Lagazio 2011):

$$E_W = -\frac{\alpha}{2} \sum_{j=1}^{W} w_j^2 \qquad (3.3)$$

Here, α is the prior contribution to the regularization error and W is the
number of network weights. This regularization parameter penalizes weights of
large magnitudes (Bishop 1995; Tibshirani 1996; Marwala 2009; Marwala and
Lagazio 2011). To solve for the weights in Eq. 3.1, the back-propagation technique
described in the next section was used.

By merging Eqs. 3.2 and 3.3, the complete objective function can be written as
follows (Bishop 1995):

$$\begin{aligned} E &= \beta E_D + \alpha E_W \\ &= -\beta \sum_{n=1}^{N} \sum_{k=1}^{K} \{t_{nk} - y_{nk}\}^2 - \frac{\alpha}{2} \sum_{j=1}^{W} w_j^2 \end{aligned} \qquad (3.4)$$

3.2.1.1   Back-Propagation Method

As described by Marwala and Lagazio (2011), the back-propagation technique is a procedure of training neural networks (Bryson and Ho 1989; Rumelhart et al. 1986; Russell and Norvig 1995). Back-propagation is a supervised learning procedure and is basically an application of the Delta rule (Rumelhart et al. 1986; Russell and Norvig 1995). Back-propagation requires that the activation function (seen in Eq. 3.1) can be differentiated and is divided into the propagation and weight update. According to Bishop (1995) the propagation characteristic has these steps:

- the propagation's output activations is enacted by forward propagating the training pattern's input into the neural network,
- the deltas of all output and hidden neurons are estimated by back-propagating the output activations in the neural network by applying the training target.

The weight update characteristic applies the output delta and input activation to get the gradient of the weight. The weight in the reverse direction of the gradient is utilized by subtracting a proportion of it from the weight and then this proportion influences the performance of the learning process. The sign of the gradient of a weight defines the direction where the error increases and this is the motive that the weight is updated in the opposite direction (Bishop 1995), and this process is recurred until convergence.

In basic terms, back-propagation is used to identify the network weights given the training data, using an optimization technique. Generally, the weights can be identified using the following iterative technique (Werbos 1974; Zhao et al. 2010; Marwala and Lagazio 2011):

$$\{w\}_{i+1} = \{w\}_i - \eta \frac{\partial E}{\partial \{w\}} (\{w\}_i) \tag{3.5}$$

In Eq. 3.5, the parameter η is the learning rate while $\{\}$ characterizes a vector. The minimization of the objective function, E, is attained by computing the derivative of the errors in Eq. 3.2 with respect to the network's weight. The derivative of the error is calculated with respect to the weight which joins the hidden layer to the output layer and may be written as follows, using the chain rule (Bishop 1995):

$$\begin{aligned}
\frac{\partial E}{\partial w_{kj}} &= \frac{\partial E}{\partial a_k} \frac{\partial a_k}{\partial w_{kj}} \\
&= \frac{\partial E}{\partial y_k} \frac{\partial y_k}{\partial a_k} \frac{\partial a_k}{\partial w_{kj}} \\
&= \sum_n f'_{outer}(a_k) \frac{\partial E}{\partial y_{nk}} z_j
\end{aligned} \tag{3.6}$$

In Eq. 3.6, $z_j = f_{inner}(a_j)$ and $a_k = \sum_{j=0}^{M} w_{kj}^{(2)} y_j$. The derivative of the error with respect to weight which connects the hidden layer to the output layer may also be written using the chain rule (Bishop 1995):

$$\frac{\partial E}{\partial w_{kj}} = \frac{\partial E}{\partial a_k} \frac{\partial a_k}{\partial w_{kj}}$$

$$= \sum_{n} f'_{inner}(a_j) \sum_{k} w_{kj} f'_{outer}(a_k) \frac{\partial E}{\partial y_{nk}} \qquad (3.7)$$

In Eq. 3.7, $a_j = \sum_{i=1}^{d} w_{ji}^{(1)} x_i$. The derivative of the objective function in Eq. 3.2 may thus be written as (Bishop 1995):

$$\frac{\partial E}{\partial y_{nk}} = (t_{nk} - y_{nk}) \qquad (3.8)$$

while that of the hyperbolic tangent function is (Bishop 1995):

$$f'_{inner}(a_j) = \sec h^2(a_j) \qquad (3.9)$$

Now that it has been explained how to estimate the gradient of the error with respect to the network weights using back-propagation procedure, Eq. 3.6 can be applied to update the network weights by using an optimization process until some pre-defined stopping criterion is realized. If the learning rate in Eq. 3.5 is fixed, then this is called the *steepest descent optimization* technique (Robbins and Monro 1951). Conversely, the steepest descent technique is not computationally efficient and, hence, an improved scheme requires to be identified. In this chapter, the scaled conjugate gradient technique was applied (Møller 1993), the subject of the next section.

3.2.1.2 Scaled Conjugate Gradient Method

The method in which the network weights are approximated from the data is by using non-linear optimization technique (Mordecai 2003). In this chapter, the scaled conjugate gradient technique (Møller 1993) was used. As described earlier in this chapter, the weight vector, which provides the minimum error, is calculated by simulating sequential steps through the weight space as presented in Eq. 3.7 until a pre-determined stopping criterion is achieved. Different procedures select this learning rate in a different ways. In this section, the gradient descent technique is described, and after that how it is modified to the conjugate gradient technique (Hestenes and Stiefel 1952). For the gradient descent technique, the step size is defined as $-\eta \partial E / \partial w$, where the parameter η is the learning rate and the gradient of the error is estimated using the back-propagation method.

If the learning rate is adequately small, the value of the error decreases at each following step until a minimum value for the error between the model prediction and training target data is achieved. The drawback with this method is that it is computationally expensive when compared to other procedures. For the conjugate gradient method, the quadratic function of the error is minimized at every step over a gradually increasing linear vector space that comprises the global minimum of the error (Luenberger 1984; Fletcher 1987; Bertsekas 1995; Marwala 2012). In the conjugate gradient technique, the following steps are used (Haykin 1999; Marwala 2009; Babaie-Kafaki et al. 2010; Marwala and Lagazio 2011):

1. Select the initial weight vector $\{w\}_0$.
2. Estimate the gradient vector $\frac{\partial E}{\partial \{w\}}(\{w\}_0)$.
3. At each step, n, apply the line search to identify $\eta(n)$ that minimizes $E(\eta)$ indicating the objective function expressed in terms of η for fixed values of w and $-\frac{\partial E}{\partial \{w\}}(\{w_n\})$.
4. Evaluate that the Euclidean norm of the vector $-\frac{\partial E}{\partial w}(\{w_n\})$ is sufficiently less than that of $-\frac{\partial E}{\partial w}(\{w_0\})$.
5. Change the weight vector using Eq. 3.4.
6. For w_{n+1}, calculate the changed gradient $\frac{\partial E}{\partial \{w\}}(\{w\}_{n+1})$.
7. Apply the Polak-Ribiére technique to estimate:

$$\beta(n+1) = \frac{\nabla E(\{w\}_{n+1})^T (\nabla E(\{w\}_{n+1}) - \nabla E(\{w\}_n)))}{\nabla E(\{w\}_n)^T \nabla E(\{w\}_n)}.$$

8. Change the direction vector

$$\frac{\partial E}{\partial \{w\}}(\{w\}_{n+2}) = \frac{\partial E}{\partial \{w\}}(\{w\}_{n+1}) - \beta(n+1)\frac{\partial E}{\partial \{w\}}(\{w\}_n).$$

9. Let $n = n + 1$ and return to step 3.
10. Terminate when the following criterion is met:

$$\varepsilon = \frac{\partial E}{\partial \{w\}}(\{w\}_{n+2}) - \frac{\partial E}{\partial \{w\}}(\{w\}_{n+1}) \quad \text{where} \ll 0.$$

The scaled conjugate gradient technique is different from the conjugate gradient scheme because it does not include the line search referred to in Step 3. The step-size is estimated using the following formula (Møller 1993):

$$\eta(n) = 2\left(\eta(n) - \left(\frac{\partial E(n)}{\partial \{w\}}(n)\right)^T H(n) \left(\frac{\partial E(n)}{\partial \{w\}}(n)\right)\right.$$

$$\left. + \eta(n)\left\|\left(\frac{\partial E(n)}{\partial \{w\}}(n)\right)\right\|^2 \Big/ \left\|\left(\frac{\partial E(n)}{\partial \{w\}}(n)\right)\right\|^2\right) \tag{3.10}$$

where H is the Hessian of the gradient. The scaled conjugate gradient technique is applied because it has been observed to resolve the optimization problems of training an MLP network more computationally effective than the gradient descent and conjugate gradient approaches (Bishop 1995).

3.2.2 Bayesian Framework

Multi-layered neural networks are parameterized classification models that make probabilistic assumptions about the data. The probabilistic outlook of these models is enabled by the application of the Bayesian framework (Marwala 2012). Learning algorithms are viewed as approaches for identifying parameter values that look probable in the light of the presented data. The learning process is performed by dividing the data into training, validation and testing sets. This is done for model selection and to ensure that the trained network is not biased towards the training data it has seen. Another way of realizing this is by the application of the regularization framework, which comes naturally from the Bayesian formulation and is now explained in detail in this chapter.

Thomas Bayes was the originator of the Bayes' theorem and Pierre-Simon Laplace generalized the theorem and it has been applied it to problems such as in engineering, statistics, political science and reliability (Stigler 1986; Fienberg 2006; Bernardo 2005; Marwala and Lagazio 2011). Initially, the Bayesian method applied uniform priors and was known as the "inverse probability" and later succeeded by a process called "frequentist statistics" also called the maximum-likelihood method. The maximum-likelihood method is intended to identify the most probable solution without concern to the probability distribution of that solution. The maximum-likelihood method technique is basically a special case of Bayesian results indicating the most probable solution in the distribution of the posterior probability function. Bayesian procedure consists of the following concepts (Bishop 1995):

- The practice of hierarchical models and the marginalization over the values of irrelevant parameters using methods such as the Markov chain Monte Carlo techniques.
- The iterative use of the Bayes' theorem as data points are acquired and after approximating a posterior distribution, the posterior equals the next prior.
- In the maximum-likelihood method, a hypothesis is a proposition which must be proven right or wrong while in a Bayesian procedure, a hypothesis has a probability.

The Bayesian method has been applied to many complex problems, including those of finite element model updating (Marwala and Sibisi 2005), missing data estimation (Marwala 2009), health risk assessment (Goulding et al. 2012), astronomy (Petremand et al. 2012), classification of file system activity (Khan 2012), simulating ecosystem metabolism (Shen and Sun 2012) and in image processing

(Thon et al. 2012). The problem of identifying the weights (w_i) and biases (with subscripts 0 in Fig. 3.1) in the hidden layers may be posed in the Bayesian form as (Box and Tiao 1973; Marwala 2012):

$$P(\{w\}|[D]) = \frac{P([D]|\{w\}) P(\{w\})}{P([D])} \qquad (3.11)$$

where $P(w)$ is the probability distribution function of the weight-space in the absence of any data, also known as the prior distribution and $D \equiv (y_1, \ldots, y_N)$ is a matrix containing the data. The quantity $P(w|D)$ is the posterior probability distribution after the data have been seen and $P(D|w)$ is the likelihood function.

3.2.2.1 Likelihood Function

The likelihood function is the notion that expresses the probability of the model which depends on the weight parameters of a model to be true. It is fundamentally the probability of the observed data, given the free parameters of the model. The likelihood can be expressed mathematically as follows, by using the sum of squares error (Edwards 1972; Bishop 1995; Marwala 2012):

$$P([D]|\{w\}) = \frac{1}{Z_D} \exp(-\beta E_D)$$

$$= \frac{1}{Z_D} \exp\left(\beta \sum_n^N \sum_k^K \{t_{nk} - y_{nk}\}^2\right) \qquad (3.12)$$

In Eq. 3.12, E_D is the sum of squares of error function, β represents the hyper-parameters, and Z_D is a normalization constant which can be approximated as follows (Bishop 1995):

$$Z_D = \int_{-\infty}^{\infty} \exp\left(\beta \sum_n^N \sum_k^K \{t_{nk} - y_{nk}\}^2\right) d\{w\} \qquad (3.13)$$

3.2.2.2 Prior Function

The *prior* probability distribution is the assumed probability of the free parameters and is approximated by a knowledgeable expert (Jaynes 1968; Bernardo 1979; Marwala and Lagazio 2011). There are many types of priors and these comprise informative and uninformative priors. An *informative prior* expresses accurate, particular information about a variable while an *uninformative prior* states general information about a variable. A prior distribution that assumes that model parameters are of the same order of magnitude can be written as follows (Bishop 1995):

$$P\left(\{w\}\right) = \frac{1}{Z_w} \exp\left(-E_W\right)$$

$$= \frac{1}{Z_w} \exp\left(-\frac{\alpha}{2} \sum_j^W w_j^2\right) \tag{3.14}$$

Parameter α represents the hyper-parameters, and Z_W is the normalization constant which can be approximated as follows (Bishop 1995):

$$Z_w = \int_{-\infty}^{\infty} \exp\left(-\frac{\alpha}{2} \sum_j^W w_j^2\right) d\{w\} \tag{3.15}$$

The prior distribution of a Bayesian method is the regularization parameter in Eq. 3.2. Regularization includes presenting additional information to the objective function, through a penalty function to solve an ill-posed problem or to prevent over-fitting to guarantee the smoothness of the solution to balance complexity with accuracy.

3.2.2.3 Posterior Function

The posterior probability is the probability of the network weights given the observed data. It is a conditional probability assigned after the appropriate evidence is taken into account (Lee 2004). It is estimated by multiplying the likelihood function with the prior function and dividing it by a normalization function. By combining Eqs. 3.11 and 3.14, the posterior distribution can be expressed as follows (Bishop 1995):

$$P\left(w|D\right) = \frac{1}{Z_s} \exp\left(\beta \sum_n^N \sum_k^K \{t_{nk} - y_{nk}\}^2 - \frac{\alpha}{2} \sum_j^W w_j^2\right) \tag{3.16}$$

where

$$Z_E(\alpha, \beta) = \int \exp\left(-\beta E_D - \alpha E_W\right) dw$$

$$= \left(\frac{2\pi}{\beta}\right)^{\frac{N}{2}} + \left(\frac{2\pi}{\alpha}\right)^{\frac{W}{2}} \tag{3.17}$$

Training the network using a Bayesian method gives the probability distribution of the weights shown in Eq. 3.1. The Bayesian method penalizes highly complex models and can choose an optimal model (Bishop 1995).

3.2.3 Automatic Relevance Determination

As described by Marwala and Lagazio (2011), an automatic relevance determination method is built by associating the hyper-parameters of the prior with each input variable. This, consequently, necessitates Eq. 3.14 to be generalized to form (MacKay 1991, 1992):

$$E_W = \frac{1}{2} \sum_k \alpha_k \{w\}^T [I_k] \{w\} \tag{3.18}$$

Here, superscript T is the transpose, k is the weight group and $[I]$ is the identity matrix. By using the generalized prior in Eq. 3.18, the posterior probability in Eq. 3.16 becomes (Bishop 1995):

$$P(\{w\}|[D], H_i) = \frac{1}{Z_s} \exp\left(\beta \sum_n \{t_n - y(\{x\}_n\}^2 - \frac{1}{2} \sum_k \alpha_k \{w\}^T [I_k] \{w\} \right)$$

$$= \frac{1}{Z_E} \exp(-E(\{w\})) \tag{3.19}$$

where

$$Z_E(\alpha, \beta) = \left(\frac{2\pi}{\beta} \right)^{\frac{N}{2}} + \prod_k \left(\frac{2\pi}{\alpha_k} \right)^{\frac{W_k}{2}} \tag{3.20}$$

Here, W_k is the number of weights in group k. The evidence can be written as follows (Bishop 1995):

$$p([D]|\alpha, \beta) = \frac{1}{Z_D Z_W} \int \exp(-E(\{w\})) \, d\{w\}$$

$$= \frac{Z_E}{Z_D Z_W}$$

$$= \frac{\left(\frac{2\pi}{\beta} \right)^{\frac{N}{2}} + \prod_k \left(\frac{2\pi}{\alpha_k} \right)^{W_k/2}}{\left(\frac{2\pi}{\beta} \right)^{\frac{N}{2}} \prod_k \left(\frac{2\pi}{\alpha_k} \right)^{W_k/2}} \tag{3.21}$$

The simultaneous estimation of the network weights and the hyper-parameters can be achieved using a number of ways including the use of Monte Carlo methods or any of its derivatives to maximize the posterior probability distribution. Another way of achieving this goal is to first maximize the log evidence and thus giving the following estimations for the hyper-parameters (Bishop 1995):

$$\beta^{MP} = \frac{N - \gamma}{2E_D\left(\{w\}^{MP}\right)} \quad (3.22)$$

$$\alpha_k^{MP} = \frac{\gamma_k}{2E_{W_k}\left(\{w\}^{MP}\right)} \quad (3.23)$$

where $\gamma = \sum_k \gamma_k$, $2E_{W_k} = \{w\}^T [I_k] \{w\}$ and

$$\gamma_k = \sum_j \left(\frac{\pi_j - \alpha_k}{\eta_j}\left([V]^T [I_k][V]\right)_{jj}\right) \quad (3.24)$$

and $\{w\}^{MP}$ is the weight vector at the maximum point and this is identified in this chapter using the scaled conjugate gradient method, η_j are the eigenvalues of $[A]$, and $[V]$ are the eigenvalues such that $[V]^T[V] = [I]$. To estimate the relevance of each input variable, the α_k^{MP}, β^{MP}, and the following steps are followed (MacKay 1991):

1. Randomly choose the initial values for the hyper-parameters.
2. Train the neural network using the scaled conjugate gradient algorithm to minimize the objective function in Eq. 3.4 and thus identify $\{w\}^{MP}$.
3. Apply the evidence framework to estimate the hyper-parameters using Eqs. 3.22 and 3.23.
4. If not converged go to Step 2.

3.3   Applications of ARD in Inflation Modeling

In this chapter we apply the ARD to identify variables that drive inflation. Inflation is measured using a concept called Consumer Price Index (CPI). Artificial intelligence has been used in the past to model inflation. For example, Şahin et al. (2004) applied neural networks and cognitive mapping to model Turkey's inflation dynamics while Anderson et al. (2012) applied neural network to estimate the functional relationship between certain component sub-indexes and the CPI extracted decision rules from the network.

Binner et al. (2010) studied the influence of money on inflation forecasting. They applied recurrent neural networks and kernel recursive least squares regression to identify the best fitting U.S.A. inflation prediction models and compared these to a naïve random walk model. Their results demonstrated no correlation between monetary aggregates and inflation. McAdam and McNelis (2005) used neural networks and models based on Phillips-curve formulations to forecast inflation. This proposed models outperformed the best performing linear models. Cao et al. (2012)

applied linear autoregressive moving average model (ARMA) and neural networks for forecasting medical cost inflation rates. The results showed that the neural network model outperformed the ARMA.

Nakamura (2005) applied neural networks for predicting inflation and the results from a U.S.A. data demonstrated that neural networks outperformed univariate autoregressive model, while Binner et al. (2006) used a neural network and a Markov switching autoregressive (MS-AR) model to predict U.S.A. inflation and found that MS-AR model performed better than neural networks.

The CPI is a measure of inflation in an economy. It measures the changes in prices of a fixed pre-selected basket of goods. A basket of goods which is used for calculating the CPI in South Africa is as follows (Anonymous 2012):

1. Food and non-alcoholic beverages: bread and cereals, meat, fish, milk, cheese, eggs, oils, fats, fruit, vegetables, sugar, sweets, desserts, and other foods
2. Alcoholic beverages and tobacco
3. Clothing and footwear
4. Housing and utilities: rents, maintenance, water, electricity, and others
5. Household contents, equipment, and maintenance
6. Health: medical equipment, outpatient, and medical service
7. Transport
8. Communication
9. Recreation and culture
10. Education
11. Restaurants and hotels
12. Miscellaneous goods and services: personal care, insurance, and financial services.

This basket is weighed and the variation of prices of these goods is tracked from month to month and this is a basis for calculating inflation. It must be noted that there is normally a debate as to whether this basket of goods is appropriate. For example, in South Africa where there are two economies, one developed and formal and another informal and under-developed, there is always a debate on the validity of the CPI. This is even more important because the salary negotiations are based on the CPI.

In this chapter, we use the CPI data from 1992 to 2011 to model the relationship between economic variables and the CPI. These economic variables are listed in Table 3.1. They represent the performance of various aspect of the economy represented by 23 variables in the agriculture, manufacturing, mining, energy, construction, etc. A multi-layered perceptron neural network with 23 input variables, 12 hidden nodes, and 1 output representing the CPI is constructed. The ARD based MLP network is trained using the scaled conjugate gradient method and all these techniques were described earlier in the chapter. The results indicating the relevance of each variable is indicated in Table 3.1.

From Table 3.1, the following variables are deemed to be essential for modeling the CPI and these are mining, transport, storage and communication, financial

Table 3.1 Automatic relevance with multi-layer perceptron and scaled conjugate gradient

Variable	Alpha	Inverse alpha	Relative weights
Agriculture fish forestry	14.4832	0.0690	1.79
Mining	5.7440	0.1741	4.51
Manufacturing	24.2071	0.0413	1.07
Electricity gas water	6.8551	0.1459	3.78
Construction	7.3717	0.1357	3.51
Retail and trade	15.0679	0.0664	1.72
Transport, storage and communication	2.3174	0.4315	11.18
Financial intermediation, insurance, real estate, and business services	0.9391	1.0648	27.59
Community, social, and personal services	0.4626	2.1616	56.00
Government services	7.2632	0.1377	3.57
Gross value added at basic prices	4.7935	0.2086	5.40
Taxes less subsidies on products	0.6467	1.5462	40.06
Affordability	1.0664	0.9377	24.29
Economic growth	4.0215	0.2487	6.44
Rand/USD exchange	25.8858	0.0386	1.00
Prime interest	5.5639	0.1797	4.66
Repo rate	5.5639	0.1797	4.66
Gross domestic product	0.2545	3.9287	101.78
Household consumption	0.4407	2.2692	58.79
Investment	0.5909	1.6924	43.84
Government consumption	7.5703	0.1321	3.42
Exports	20.8664	0.0479	1.24
Imports	5.9678	0.0386	1.00

intermediation, insurance, real estate and business services, community, social and personal services, gross value added at basic prices, taxes less subsidies on products, affordability, economic growth, repo rate, gross domestic product, household consumption, and investment.

It should be noted, however, that these results are purely based on the data set that was analyzed and the methodology that was used which is the ARD that is based on the MLP. These conclusions may change from one economy to another and from one methodology to another e.g. support vector machines instead of the MLP.

3.4   Conclusions

This chapter presented the Bayesian and the evidence frameworks to create an automatic relevance determination technique. The ARD method was then applied to determine the relevance of economic variables that are essential for driving the consumer price index. It is concluded that for the data analyzed using the MLP based ARD technique, the variables driving the CPI are mining, transport, storage and communication, financial intermediation, insurance, real estate and business services, community, social and personal services, gross value added at basic prices, taxes less subsidies on products, affordability, economic growth, repo rate, gross domestic product, household consumption, and investment.

References

Alsharafat W (2013) Applying artificial neural network and extended classifier system for network intrusion detection. Int Arab J Inf Technol 10:art. no. 6-3011

Anderson RG, Binner JM, Schmidt VA (2012) Connectionist-based rules describing the pass-through of individual goods prices into trend inflation in the United States. Econ Lett 117:174–177

Anonymous (2012) CPI data http://www.statssa.gov.za/. Last accessed 03 Sept 2012

Babaie-Kafaki S, Ghanbari R, Mahdavi-Amiri N (2010) Two new conjugate gradient methods based on modified secant equations. J Comput Appl Math 234:1374–1386

Bernardo JM (1979) Reference posterior distributions for Bayesian inference. J R Stat Soc 41:113–147

Bernardo JM (2005) Reference analysis. Handb Stat 25:17–90

Bertsekas DP (1995) Non-linear programming. Athenas Scientific, Belmont

Binner JM, Elger CT, Nilsson B, Tepper JA (2006) Predictable non-linearities in U.S. inflation. Econ Lett 93:323–328

Binner JM, Tino P, Tepper J, Anderson R, Jones B, Kendall G (2010) Does money matter in inflation forecasting? Physica A Stat Mech Its Appl 389:4793–4808

Bishop CM (1995) Neural networks for pattern recognition. Oxford University Press, Oxford

Böck M, Ogishima S, Tanaka H, Kramer S, Kaderali L (2012) Hub-centered gene network reconstruction using automatic relevance determination. PLoS ONE 7:art. no. e35077

Box GEP, Tiao GC (1973) Bayesian inference in statistical analysis. Wiley, Hoboken

Browne A, Jakary A, Vinogradov S, Fu Y, Deicken RF (2008) Automatic relevance determination for identifying thalamic regions implicated in schizophrenia. IEEE Trans Neural Netw 19:1101–1107

Bryson AE, Ho YC (1989) Applied optimal control: optimization, estimation, and control. Xerox College Publishing, Kentucky

Cao Q, Ewing BT, Thompson MA (2012) Forecasting medical cost inflation rates: a model comparison approach. Decis Support Syst 53:154–160

Cybenko G (1989) Approximations by superpositions of sigmoidal functions. Math Control Signal Syst 2:303–314

Edwards AWF (1972) Likelihood. Cambridge University Press, Cambridge

Fienberg SE (2006) When did Bayesian inference become "Bayesian"? Bayesian Anal 1:1–40

Fletcher R (1987) Practical methods of optimization. Wiley, New York

Fu Y, Browne A (2008) Investigating the influence of feature correlations on automatic relevance determination. In: Proceedings of the international joint conference on neural Networks, Hong Kong, 2008, pp 661–665

Goulding R, Jayasuriya N, Horan E (2012) A Bayesian network model to assess the public health risk associated with wet weather sewer overflows discharging into waterways. Water Res 46:4933–4940

Hassoun MH (1995) Fundamentals of artificial neural networks. MIT Press, Cambridge, MA

Haykin S (1999) Neural networks. Prentice-Hall, Upper Saddle River

Hestenes MR, Stiefel E (1952) Methods of conjugate gradients for solving linear systems. J Res Nat Bur Stand 6:409–436

Huang Y, Beck JL, Wu S, Li H (2012) Stochastic optimization using automatic relevance determination prior model for Bayesian compressive sensing. In: Proceedings of SPIE – the international society for optical engineering, San Diego, 2012, art. no. 834837

Hurwitz E, Marwala T (2011) Suitability of using technical indicators as potential strategies within intelligent trading systems. In: Proceedings of the IEEE international conference on systems, man, and cybernetics, Anchorage, 2011, pp 80–84

Ikuta C, Uwate Y, Nishio Y (2012) Multi-layer perceptron with positive and negative pulse glial chain for solving two-spirals problem. In: Proceedings of the international joint conference on neural networks, Brisbane, 2012, art. no. 6252725

Jacobs JP (2012) Bayesian support vector regression with automatic relevance determination kernel for modeling of antenna input characteristics. IEEE Trans Antenna Propag 60:2114–2118

Jaynes ET (1968) Prior probabilities. IEEE Trans Syst Sci Cybern 4:227–241

Khan MNA (2012) Performance analysis of Bayesian networks and neural networks in classification of file system activities. Comput Secur 31:391–401

Khoza M, Marwala T (2012) Computational intelligence techniques for modelling an economic system. In: Proceedings of the international joint conference on neural networks, Brisbane, 2012, pp 1–5

Lee PM (2004) Bayesian statistics, an introduction. Wiley, Hoboken

Leke B, Marwala T (2005) Optimization of the stock market input time-window using Bayesian neural networks. In: Proceedings of the IEEE international conference on service operations, logistics and informatics, Beijing, 2005, pp 883–894

Leke B, Marwala T, Tettey T (2007) Using inverse neural network for HIV adaptive control. Int J Comput Intell Res 3:11–15

Li X, Li L (2012) IP core based hardware implementation of multi-layer perceptrons on FPGAs: a parallel approach. Adv Mater Res 433–440:5647–5653

Lisboa PJG, Etchells TA, Jarman IH, Arsene CTC, Aung MSH, Eleuteri A, Taktak AFG, Ambrogi F, Boracchi P, Biganzoli E (2009) Partial logistic artificial neural network for competing risks regularized with automatic relevance determination. IEEE Trans Neural Netw 20:1403–1416

Luenberger DG (1984) Linear and non-linear programming. Addison-Wesley, Reading

Lunga D, Marwala T (2006) Online forecasting of stock market movement direction using the improved incremental algorithm. Lect Note Comput Sci 4234:440–449

MacKay DJC (1991) Bayesian methods for adaptive models. Ph.D. thesis, California Institute of Technology, Pasadena

MacKay DJC (1992) A practical Bayesian framework for back propagation networks. Neural Comput 4:448–472

Martínez-Rego D, Fontenla-Romero O, Alonso-Betanzos A (2012) Nonlinear single layer neural network training algorithm for incremental, nonstationary and distributed learning scenarios. Pattern Recognit 45:4536–4546

Marwala T (2000) On damage identification using a committee of neural networks. J Eng Mech 126:43–50

Marwala T (2001) Probabilistic fault identification using a committee of neural networks and vibration data. J Aircr 38:138–146

Marwala T (2003) Fault classification using pseudo modal energies and neural networks. Am Inst Aeronaut Astronaut J 41:82–89

Marwala T (2009) Computational intelligence for missing data imputation, estimation and management: knowledge optimization techniques. IGI Global Publications, New York

Marwala T (2012) Condition monitoring using computational intelligence methods. Springer, London

Marwala T, Hunt HEM (1999) Fault identification using finite element models and neural networks. Mech Syst Signal Process 13:475–490

Marwala T, Lagazio M (2011) Militarized conflict modeling using computational intelligence techniques. Springer, London

Marwala T, Sibisi S (2005) Finite element model updating using Bayesian framework and modal properties. J Aircr 42:275–278

McAdam P, McNelis P (2005) Forecasting inflation with thick models and neural networks. Econ Model 22:848–867

Meena K, Subramaniam K, Gomathy M (2013) Gender classification in speech recognition using fuzzy logic and neural network. Int Arab J Inf Technol 10:art. no. 4476-7

Mohamed N, Rubin D, Marwala T (2006) Detection of epileptiform activity in human EEG signals using Bayesian neural networks. Neural Inf Process Lett Rev 10:1–10

Møller AF (1993) A scaled conjugate gradient algorithm for fast supervised learning. Neural Netw 6:525–533

Mordecai A (2003) Non-linear programming: analysis and methods. Dover Publishing, New York

Mørupa M, Hansena LK (2009) Automatic relevance determination for multi-way models. J Chemom 23:352–363

Nakamura E (2005) Inflation forecasting using a neural network. Econ Lett 86:1–8

Nasir AA, Mashor MY, Hassan R (2013) Classification of acute leukaemia cells using multilayer perceptron and simplified fuzzy ARTMAP neural networks. Int Arab J inf Technol 10:art. no. 4626-12

Nummenmaa A, Auranen T, Hämäläinen MS, Jääskeläinen IP, Sams M, Vehtari A, Lampinen J (2007) Automatic relevance determination based hierarchical Bayesian MEG inversion in practice. Neuroimage 37:876–889

Oh CK, Beck JL, Yamada M (2008) Bayesian learning using automatic relevance determination prior with an application to earthquake early warning. J Eng Mech 134:1013–1020

Patel P, Marwala T (2006) Neural networks, fuzzy inference systems and adaptive-neuro fuzzy inference systems for financial decision making. Lect Note Comput Sci 4234:430–439

Petremand M, Jalobeanu A, Collet C (2012) Optimal bayesian fusion of large hyperspectral astronomical observations. Stat Methodol 9:1572–3127

Prakash G, Kulkarni M, Sripati Acharya U, Kalyanpur MN (2012) Classification of FSO channel models using radial basis function neural networks and their performance with luby transform codes. Int J Artif Intell 9:67–75

Rezaeian-Zadeh M, Zand-Parsa S, Abghari H, Zolghadr M, Singh VP (2012) Hourly air temperature driven using multi-layer perceptron and radial basis function networks in arid and semi-arid regions. Theor Appl Climatol 109:519–528

Robbins H, Monro S (1951) A stochastic approximation method. Ann Math Stat 22:400–407

Rosenblatt F (1961) Principles of neurodynamics: perceptrons and the theory of brain mechanisms. Spartan, Washington DC

Rumelhart DE, Hinton GE, Williams RJ (1986) Parallel distributed processing: explorations in the microstructure of cognition. MIT Press, Cambridge, MA

Russell S, Norvig P (1995) Artificial intelligence: a modern approach. Prentice Hall, Englewood Cliffs

Russell MJ, Rubin DM, Wigdorowitz B, Marwala T (2008) The artificial larynx: a review of current technology and a proposal for future development. Proc Int Fed Med Biol Eng 20:160–163

Russell MJ, Rubin DM, Wigdorowitz B, Marwala T (2009a) Pattern recognition and feature selection for the development of a new artificial larynx. In: Proceedings of the 11th world congress on medical physics and biomedical engineering, Munich, 2009, pp 736–739

Russell MJ, Rubin DM, Marwala T, Wigdorowitz B (2009b) A voting and predictive neural network system for use in a new artificial larynx. Proc IEEE ICBPE. doi:10.1109/ICBPE.2009.5384105

Şahin ŞÖ, Ülengin FN, Ülengin B (2004) Using neural networks and cognitive mapping in scenario analysis: the case of Turkey's inflation dynamics. Eur J Oper Res 158:124–145

Sanz J, Perera R, Huerta C (2012) Gear dynamics monitoring using discrete wavelet transformation and multi-layer perceptron neural networks. Appl Soft Comput J 12:2867–2878

Shaltaf S, Mohammad A (2013) A hybrid neural network and maximum likelihood based estimation of chirp signal parameters. Int Arab J Inf Technol 10:art. no. 4580-12

Shen X, Sun T (2012) Applications of bayesian modeling to simulate ecosystem metabolism in response to hydrologic alteration and climate change in the Yellow River Estuary, China. Procedia Environ Sci 13:790–796

Shutin D, Buchgraber T, Kulkarni SR, Poor HV (2011) Fast variational sparse Bayesian learning with automatic relevance determination for superimposed signals. IEEE Trans Signal Process 59:6257–6261

Shutin D, Kulkarni SR, Poor HV (2012) Incremental reformulated automatic relevance determination. IEEE Trans Signal Process 60:4977–4981

Sinha K, Chowdhury S, Saha PD, Datta S (2013) Modeling of microwave-assisted extraction of natural dye from seeds of *Bixa orellana* (Annatto) using response surface methodology (RSM) and artificial neural network (ANN). Ind Crop Prod 41:165–171

Smyrnakis MG, Evans DJ (2007) Classifying ischemic events using a Bayesian inference multilayer perceptron and input variable evaluation using automatic relevance determination. Comput Cardiol 34:305–308

Stigler SM (1986) The history of statistics. Harvard University Press, Cambridge, MA

Taspinar N, Isik Y (2013) Multiuser detection with neural network MAI detector in CDMA systems for AWGN and Rayleigh fading asynchronous channels. Int Arab J Inf Technol 10:art. no. 4525-5

Thon K, Rue H, Skrøvseth SO, Godtliebsen F (2012) Bayesian multiscale analysis of images modeled as Gaussian Markov random fields. Comput Stat Data Anal 56:49–61

Tibshirani R (1996) Regression shrinkage and selection via the lasso. J R Stat Soc 58:267–288

Ulusoy I, Bishop CM (2006) Automatic relevance determination for the estimation of relevant features for object recognition. In: Proceedings of the IEEE 14th signal processing and communication applications, Antalya, 2006, pp 1–4

Valdés JJ, Romero E, Barton AJ (2012) Data and knowledge visualization with virtual reality spaces, neural networks and rough sets: application to cancer and geophysical prospecting data. Expert Syst Appl 39:13193–13201

Van Calster B, Timmerman D, Nabney IT, Valentin L, Van Holsbeke C, Van Huffel S (2006) Classifying ovarian tumors using Bayesian multi-layer perceptrons and automatic relevance determination: a multi-center study. Proc Eng Med Biol Soc 1:5342–5345

Vilakazi BC, Marwala T (2007) Condition monitoring using computational intelligence. In: Laha D, Mandal P (eds) Handbook on computational intelligence in manufacturing and production management, illustrated edn. IGI Publishers, New York

Wang D, Lu WZ (2006) Interval estimation of urban ozone level and selection of influential factors by employing automatic relevance determination model. Chemosphere 62:1600–1611

Werbos PJ (1974) Beyond regression: new tool for prediction and analysis in the behavioral sciences. Ph.D. thesis, Harvard University, Cambridge

Wu D (2012) An improved multi-layer perceptron neural network for scattered point data surface reconstruction. ICIC Express Lett Part B Appl 3:41–46

Wu W, Chen Z, Gao S, Brown EN (2010) Hierarchical Bayesian modeling of inter-trial variability and variational Bayesian learning of common spatial patterns from multichannel EEG. In: Proceedings of the 2010 IEEE international conference on acoustics speech and signal processing, Dallas, 2010, pp 501–504

Yoon Y, Peterson LL (1990) Artificial neural networks: an emerging new technique. In: Proceedings of the ACM SIGBDP conference on trends and directions in expert systems, Cambridge, 1990, pp 417–422

Zhang X, Gou L, Hou B, Jiao L (2010) Gaussian process classification using automatic relevance determination for SAR target recognition. In: Proceedings of SPIE – the international society for optical engineering, art. no. 78300R

Zhao Z, Xin H, Ren Y, Guo X (2010) Application and comparison of BP neural network algorithm in MATLAB. In: Proceedings of the international conference on measurement technology and mechatron automat, New York, 2010, pp 590–593

Chapter 4
Neural Approaches to Economic Modeling

Abstract This chapter describes the multi-layered perceptron (MLP), radial basis functions (RBF), and support vector machines (SVM). These techniques are then applied to model inflation by training these networks using the maximum-likelihood method. The results indicated that the SVM gives the best results followed by the MLP and then the RBF. The SVM with an exponential RBF gave the best results.

4.1 Introduction

This chapter applies the multi-layered perceptron (MLP), radial basis functions (RBF) and support vector machines (SVM) to inflation modeling. It uses economic variables that were identified in Chap. 3 by using the automatic relevance determination method that was based on the MLP. Modeling inflation is an important undertaking for economic planning as it impacts on areas of vital economic activities.

Welfe (2000) applied a Bayesian framework and the Johansen method to model inflation in Poland, while Price and Nasim (1999) applied co-integration and the P-star approach to model inflation and money demand in Pakistan.

De Gooijer and Vidiella-i-Anguera (2003) successfully applied the multiplicative seasonal self-exciting threshold autoregressive which factored monthly inflation rates and seasonal fluctuations at the same time. Gregoriou and Kontonikas (2009) modeled non-linearity in inflation deviations from the target of various countries using an exponential smooth transition autoregressive (ESTAR) and Markov regime-switching models. They observed that the ESTAR model was able to capture the non-linear behavior of inflation. In addition they observed a fast adjustment process in countries that regularly under estimate inflation target while countries that over-estimate inflation target displayed a slower revision back to equilibrium. The ESTAR model was also observed to outperform the Markov regime-switching model.

T. Marwala, *Economic Modeling Using Artificial Intelligence Methods*, Advanced
Information and Knowledge Processing, DOI 10.1007/978-1-4471-5010-7_4,
© Springer-Verlag London 2013

Aïssa et al. (2007) modelled inflation persistence with periodicity changes in fixed and pre-determined prices models. In particular, they considered the relationship between the length of the decision interval for a specified duration of prices and the dynamic properties of inflations. They observed that higher frequency types of information, overlapping contracts and hybrid price models forecast inflation persistence.

Mehrotra et al. (2010) applied the hybrid New Keynesian Phillips Curve (NKPC) for modelling inflation in China and demonstrated that the NKPC gave good results of inflation. A probit analysis showed that the forward-looking inflation element and the output gap were vital inflation variables in provinces that that had undergone market reform.

Kukal and Van Quang (2011) applied the multi-layered perceptron and radial basis function neural networks to model how Czech National Bank handled its repo rate when it implemented its monetary policy and observed that the MLP network was better at modeling the repo rate than the RBF.

Choudhary and Haider (2012) compared neural networks to autoregressive models for inflation forecasting and found that neural networks performed better than the autoregressive models.

Düzgün (2010) compared the generalized regression neural networks to the autoregressive integrated moving average (ARIMA) and found that neural network model performed better than the ARIMA model in forecasting the Consumer Price Index.

Santana (2006) compared neural networks to exponential smoothing and ARIMA methods for forecasting Colombian Inflation and observed that neural network forecasts were more accurate than the forecasts from exponential smoothing and ARIMA methods.

Binner et al. (2005) compared linear forecasting models and neural networks for modeling Euro inflation rate. The results obtained indicated that neural networks gave better forecasts than linear models.

Moshiri and Cameron (2000) compared neural network to econometric models for forecasting inflation and they observed that neural networks were able to forecast marginally better than traditional econometric methods.

Other approaches that have been used to model inflation are support vector machines (Wang et al. 2012), genetic optimization (Huang 2012a), neural networks (Wang and Fan 2010; Ge and Yin 2012), dynamic factor estimation method (Kapetanios 2004), combinatorial and optimal networks (Luo and Huang 2012), and elitist neuro-genetic network model (Mitra 2012). Other aspects of inflation that have been modeled are the relationship between short-term and long term inflation expectations (Jochmann et al. 2010), pass-through of individual goods prices into trend inflation (Anderson et al. 2012), identification of post-stabilization inflation dynamics (Dimirovski and Andreeski 2006) and medical costs inflation rates (Cao et al. 2012).

From this literature review, it is evident that artificial intelligence methods are viable tools for modelling inflation. The next sections will describe multi-layered perceptrons, radial basis functions, and support vector machines which are techniques that are used to model inflation in this chapter.

4.2 Multi-layer Perceptron Neural Networks

The neural network model considered in this chapter is the multi-layer perceptron. The MLP is a feed-forward neural network model that estimates a relationship between sets of input data and a set of corresponding output. Its basis is the standard linear perceptron and it uses three or more layers of neurons, also called nodes, with non-linear activation functions. It is more effective than the perceptron because it can discriminate data that is not linearly separable or separable by a hyper-plane.

The multi-layered perceptron has been used to model many diverse complex systems in areas such as mechanical engineering (Marwala 2010, 2012), missing data (Marwala 2009), interstate conflict (Marwala and Lagazio 2011), soil water assessment (Singh et al. 2012), solid waste (Arebey et al. 2012), in biomarkers to infer characteristics of contaminant exposure (Karami et al. 2012), collision avoidance of a ship (Ahn et al. 2012), speech recognition (Ikbal et al. 2012), and power transformers fault diagnosis (Souahlia et al. 2012). More relevant to the topic of economic modeling, Tung et al. (2004) applied neural networks to build an early warning system for banking failure, while Yümlü et al. (2005) used neural networks for Istanbul stock exchange prediction. Tsai and Wu (2008) applied ensembles of multi-layer perceptrons for bankruptcy prediction and credit scoring, and observed that the ensemble performed better than the stand-alone networks.

As described in Chap. 3, the MLP neural network is made up of multiple layers of computational components typically inter-connected in a feed-forward way (Haykin 1999; Hassoun 1995; Marwala 2012). Every neuron in one layer is linked to the neurons of the following layer. A completely linked two layered MLP network is used in this chapter. A NETLAB® toolbox that runs in MATLAB® which was built by Nabney (2001) was implemented to build the MLP neural network. The network which was discussed in Chap. 3 can be mathematically described as follows (Haykin 1999):

$$
y_k = f_{outer} \left(\sum_{j=1}^{M} w_{kj}^{(2)} \, f_{inner} \left(\sum_{i=1}^{d} w_{ji}^{(1)} x_i + w_{j0}^{(1)} \right) + w_{k0}^{(2)} \right) \tag{4.1}
$$

Here $w_{ji}^{(1)}$ and $w_{ji}^{(2)}$ are weights in the first and second layer, respectively, from input i to hidden unit j, M is the number of hidden units, d is the number of output units while $w_{j0}^{(1)}$ and $w_{k0}^{(2)}$ are the weight parameters that represent the biases for the hidden unit j and the output unit k. These weight parameters can be interpreted as an instrument that ensures that the model comprehends the data. In this chapter, the parameter $f_{outer}(\bullet)$ is the linear function while f_{inner} is the hyperbolic tangent function. As described in Chap. 3, the weight vector in Eq. 4.1 is identified using the scaled conjugate gradient method that is premised on the maximum-likelihood approach (Møller 1993).

4.3 Radial-Basis Function (RBF)

The other neural network technique that is applied in this chapter is the radial basis function. The RBF neural networks are feed-forward networks which are trained using a supervised training algorithm (Haykin 1999; Buhmann and Ablowitz 2003; Marwala and Lagazio 2011).

Marcozzi et al. (2001) successfully applied RBF and finite-difference algorithms for option pricing, while Jing and Ning-qiang (2012) applied the RBF in power system's damping analysis. Guo et al. (2012) successfully applied RBF to predict financial risk of the thermal power industry. Other successful applications of the RBF to model complex problems include controlling chaotic systems (Hsu et al. 2012), braking system of a railway vehicle (Yang and Zhang 2012) and prediction of protein interaction sites (Chen et al. 2012b).

Golbabai and Milev (2012) solved a partial integro-differential equation problem with a free boundary in an American option model. They applied front-fixing transformation of the underlying asset variable to deal with the free boundary conditions. A radial basis functions technique was applied to realize a first order nonlinear equation and they used the Predictor–Corrector technique to solve the nonlinear equations. The results obtained were similar to those achieved by other numerical methods.

Yu et al. (2012) applied a hybrid of particle swarm optimization and RBF (PSO-RBF) for predicting China's primary energy demands in 2020. They analyzed the energy demand for the period from 1980 to 2009 based on the gross domestic product (GDP), population, proportion of industry in GDP, urbanization rate, and share of coal energy. The results they obtained showed that the PSO–RBF had fewer hidden nodes and smaller estimated errors than traditional neural network models.

The RBF is normally arranged with a single hidden layer of units whose activation function is chosen from a class of functions known as the basis functions. The activation of the hidden units in an RBF neural network is described by a non-linear function of the distance between the input vector and a vector representing the centers of mass of the data (Bishop 1995). Although related to back-propagation radial basis function networks have a number of advantages. They train much faster than the MLP networks and are less disposed to problems with non-stationary inputs owing to the behavior of the radial basis function (Bishop 1995). The RBF network is shown in Fig. 4.1 and is expressed mathematically as follows (Buhmann and Ablowitz 2003; Marwala and Lagazio 2011):

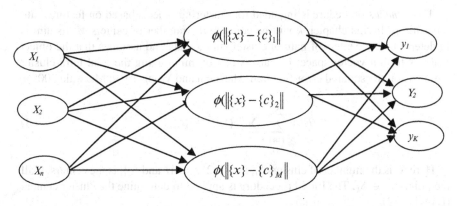

Fig. 4.1 Radial basis function network having two layers of Adaptive weights (Reprinted from Marwala and Lagazio 2011 with kind permission from Springer Science + Business Media B.V)

$$y_k\left(\{x\}\right) = \sum_{j=1}^{M} w_{jk}\phi\left(\|\{x\} - \{c\}_j\|\right) \tag{4.2}$$

Here, w_{jk} are the output weights, corresponding to the link between a hidden unit and an output unit, M denotes the number of hidden units, $\{c\}_j$ is the center for the jth neuron, $\phi\left(\{x\}\right)$ is the jth non-linear activation function, $\{x\}$ the input vector, and $k = 1, 2, 3, \ldots, M$ (Bishop 1995; Marwala and Lagazio 2011). Once more, as in the MLP, the choice of the number of hidden nodes M is part of the model selection procedure.

The activation in the hidden layers implemented in this chapter is a Gaussian distribution which can be written as follows (Bishop 1995):

$$\left(\phi\left(\|\{x\} - \{c\}\|\right) = \exp\left(-\beta(\{x\} - \{c\})^2\right)\right) \tag{4.3}$$

where β is constant.

As explained by Marwala (2009), the radial basis function is different from the multi-layered perceptron because it has weights in the outer layer only and the hidden nodes have centers. Training the radial basis function network involves identifying two sets of parameters and these are the centers and the output weights and both of these can be regarded as free parameters in a regression framework. Although the centers and network weights can both be estimated at the same time, in this chapter we apply a two stage training process to identify the centers. The first is to use self-organizing maps to estimate the centers and, in this chapter, we apply the *k-means* clustering technique (Hartigan 1975; Marwala 2009). The step of identifying the centers only considers the input space whereas the identification of the network weights uses both the input and output space.

The *k-means* procedure is intended for clustering objects based on features into *k* partitions. In this chapter, *k* will be equal to the number of centers *M*. Its aim is to determine the centers of natural clusters in the data and assumes that the object features form a vector space. It realizes this by minimizing the total intra-cluster variance, or, the squared error function (Hartigan and Wong 1979; Marwala 2009):

$$E = \sum_{i=1}^{C} \sum_{x_j \in S_i} \left(\{x\}_j - \{c\}_i \right)^2 \qquad (4.4)$$

Here, *C* is the number of clusters S_i, $i = 1, 2, \ldots, M$ and $\{c\}_i$ is the centers of all the points $x_j \in S_i$. The Lloyd procedure is applied to determine the cluster centers (Lloyd 1982).

As described by Lloyd (1982), the procedure is started by randomly segmenting the input space into *k* sets. Then the mean point is computed for each set, and then a new partition is created by linking each point with the nearest center. The centroids are then recomputed for the new clusters, and the procedure is reiterated by varying these two steps until convergence. Convergence is realized when the centroids no longer change or the points no longer switch clusters.

After the centers have been determined, the network weights then need to be estimated, given the training data. To realize this, the Moore-Penrose pseudo inverse (Moore 1920; Penrose 1955; Golub and Van Loan 1996; Marwala 2009) is applied. Once the centers have been determined, then the approximation of the network weights is a linear process (Golub and Van Loan 1996). With the training data and the centers determined, Eq. 4.2 can then be rewritten as follows (Marwala and Lagazio 2011):

$$\left[y_{ij} \right] = \left[\phi_{ik} \right] \left[w_{kj} \right] \qquad (4.5)$$

where $\left[y_{ij} \right]$ is the output matrix with *i* denoting the number of training examples, and *j* denoting the number of output instances. The parameter $[\phi_{ik}]$ is the activation function matrix in the hidden layer, with *i* denoting the training examples, and *k* denoting the number of hidden neurons. The parameter $\left[w_{kj} \right]$ is the weight matrix, with *k* denoting the number of hidden neurons and *j* denoting the number of output instances. It can thus be observed that to estimate the weight matrix $\left[w_{kj} \right]$, the activation function matrix $[\phi_{ik}]$ ought to be inverted. Nevertheless, this matrix is not square and consequently it can be inverted using the Moore-Penrose pseudo-inverse is as follows (Marwala and Lagazio 2011):

$$[\phi_{ik}]^* = \left([\phi_{ik}] [\phi_{ik}]^T \right)^{-1} [\phi_{ik}]^T \qquad (4.6)$$

This means that the weight matrix may be estimated as follows (Marwala and Lagazio 2011):

$$[w_{kj}] = [\phi_{ik}]^* [y_{ij}] \tag{4.7}$$

A NETLAB® toolbox that runs in MATLAB® and explained in (Nabney 2001) was applied to implement both the RBF and the MLP architectures.

4.3.1 Model Selection

Given that techniques of determining the weights that are learned from the training data given an RBF or even an MLP model, the following undertaking is how to select a suitable model, given by the size of the hidden nodes. The procedure of choosing suitable model is called as model selection (Burnham and Anderson 2002). The idea of deriving a model from data is a no-unique problem. This is for the reason that many models are capable of describing the training data and, consequently, it becomes difficult to determine the most suitable model. The basic method for model selection is based on two principles and these are the goodness of fit and complexity. The goodness of fit, basically, suggests that a good model should be capable of predicting the validation data which has not been used to train the networks. The complexity of the model is actually based on the principle of Occam's razor that recommends that the preferred model should be the simplest one.

There are vital issues such as:

- How do you choose a compromise between the goodness of fit and the complexity of the model?
- How are these characteristics applied?

The goodness of fit is evaluated, in this chapter, by calculating the error between the model prediction and the validation set whereas the complexity of the model is evaluated by the number of free parameters in the data. As explained by Marwala (2009), free parameters in the MLP model are defined as the network weights and biases whereas in the RBF network are defined as the network centers and weights. In this chapter, model selection is regarded as an instrument of choosing a model that has a good probability of approximating the validation data that it has not been exposed to during network training and the bias and variance are indicators of the capability of this model to function in an acceptable manner.

The MLP is trained in this chapter using the scaled conjugate gradient technique while for the RBF the *k-means* and pseudo-inverse techniques are used estimate the free parameters (centers and weights).

4.4 Support Vector Regression

The type of artificial intelligence method called support vector machines are a supervised learning technique applied principally for classification and are resulting from statistical learning theory and were first presented by Vapnik (1995, 1998). They have also been modified to handle regression problems (Gunn 1997; Chang and Tsai 2008; Chuang 2008; Hu et al. 2012; Orchel 2012). As described in Marwala (2009), Pires and Marwala (2004) successfully applied support vector machines for option pricing and additionally extended these to the Bayesian framework whereas Gidudu et al. (2007) successfully applied support vector machines for image classification.

Furthermore, Khemchandani et al. (2009) successfully applied the regularized least squares fuzzy support vector regression for financial time series forecasting. Chen et al. (2012a) successfully used support vector machines for seismic assessment of school buildings, while Golmohammadi et al. (2012) used support vector machines for quantitative structure–activity relationship estimation of blood-to-brain partitioning behavior. Zhang et al. (2006) successfully used support vector regression for on-line health monitoring of large-scale structures whereas Chen et al. (2012c) applied SVM for customer churn prediction.

One of the difficulties of support vector regression is the computational load required to train them. Guo and Zhang (2007) established methods for accelerating support vector regression whereas Üstün et al. (2007) visualized support vector regression models. Other successful applications of support vector machine include prediction of carbon monoxide concentrations (Yeganeh et al. 2012), bearing fault detection in industrial environments (Gryllias and Antoniadis 2012), defect recognition in SonicIR (Zeng et al. 2012), classification of small-grain weed species (Rumpf et al. 2012), crash injury severity analysis (Li et al. 2012) and mechanical fault condition monitoring in induction motor (Salem et al. 2012). Huang (2012b) applied genetic algorithms and support vector regression for hybrid stock selection model and demonstrated that the proposed method gave better investment returns than the benchmark.

The general notion behind support vector regression is to relate the input space to an output space. Assuming we have the training dataset with one input and one output being taken into account: $\{(x_1, y_1), \ldots, (x_i, y_i)\} \subset \chi \times \Re$, where χ is the space of the input parameters and $\Re$ represents the real number set. We intend to identify a function $f(x)$ that will relate the training inputs to the training outputs. In Support Vector regression the objective is to identify this function that has at most ε deviation from the actual training targets y_l. We can identify a number of functions $f(x)$ to relate training inputs to training outputs. These functions are called kernel functions nevertheless these cannot just be any functions because kernel functions have to satisfy to some principles (Joachims 1999). As described by many other researchers we study a linear kernel function (Lang 1987):

$$f(x) = \langle w, x \rangle + b \quad with \quad w \in \chi, \quad b \in \Re \qquad (4.8)$$

Table 4.1 Common types of kernels

Kernel Type	Equation
Linear	$k\left(x_i, x_j\right) = \left(x_i.x_j\right)$
Polynomial (inhomogeneous)	$k\left(x_i, x_j\right) = \left(x_i.x_j + c\right)^d$ where $d \geq 2$ and $c > 0$
Radial basis function	$k\left(x_i, x_j\right) = \exp\left(-\gamma \|x_i - x_j\|^2\right)$ for $\gamma > 0$
Exponential radial basis function	$k\left(x_i, x_j\right) = \exp\left(-\frac{1}{2\sigma^2}\|x_i - x_j\|^2\right)$
Spline	$k\left(x_i, x_j\right) = 1 + x_i x_j + \frac{1}{2}\|x_i - x_j\| \min\left(x_i, x_j\right)^2 + \frac{\min\left(x_i, x_j\right)^3}{3}$

where $\langle .,. \rangle$ represents the dot product. Other kernel functions that can be used include exponential radial basis function, linear spline, radial basis function and polynomial which are shown in Table 4.1.

It is intended to identify small values for w and this can be achieved by minimizing the Euclidean norm $\|w\|^2$ (Drezet and Harrison 2001). A slack variables ξ_i, ξ_i^* can be included in order to ensure that particular infeasible constraints in the minimization of the Euclidean norm can be applied and the minimization problem then becomes (Oliveira 2006):

$$\min \frac{1}{2}\|w\|^2 + C \sum_{i=1}^{l} \left(\xi_i + \xi_i^*\right) \tag{4.9}$$

$$subject\ to\ \begin{cases} y_i - \langle w, x_i \rangle - b \leq & \varepsilon + \xi_i \\ \langle w, x_i \rangle + b - y_i \leq & \varepsilon + \xi_i^* \\ \xi_i, \xi_i^* \geq & 0 \end{cases} \tag{4.10}$$

where l is the number of training points used. The constraints in Eq. 4.10 deal with an ε-insensitive loss function applied to penalize particular training points that are outside of the bound given by ε which is a value selected.

Some other loss functions that may be used include smooth non-convex loss function (Zhong 2012) and the Huber loss function which can also be applied but the most common one is the ε-insensitive loss function which is given by (Gunn 1997):

$$|\xi|_\varepsilon = \begin{cases} 0 & if\ \ |\xi| \leq \varepsilon \\ |\xi| - \varepsilon & otherwise \end{cases} \tag{4.11}$$

As explained in Marwala (2009), the value for C is the extent to which deviations from ε are accepted (Trafalis and Ince 2000). It can be viewed as a degree of over-fitting a function to the training data. If the value of C is set too high then the function $f(x)$ will be too well fitted to the training data and will not forecast well on data that

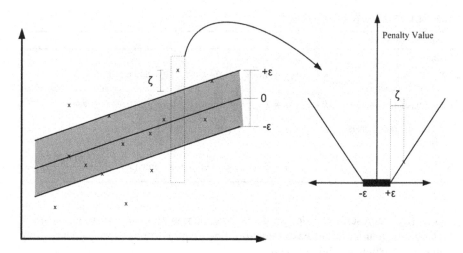

Fig. 4.2 Linear support vector regression for a set of data (*left*) and the ε-insensitive loss function (*right*) (Adapted from Gunn 1997)

has not seen by the training of the function. This implies that data located outside of the bounds given by ε are not penalized sufficiently resulting in the function being too well fitted to the training data (Trafalis and Ince 2000). An illustration of a linear function being fitted to the training data can be observed in Fig. 4.2 with the bounds displayed.

The function in Fig. 4.2 (right hand side) is applied to penalize those data points that are located outside of the bounds (left hand side). The further a data point is located outside of one of the bounds, the more the data point is penalized and, therefore, it is factored less in the estimation of the function. Those data points that are located within the bounds of the function are not penalized at all and their matching slack variable values (ξ_i, ξ_i^*) are assigned zero and accordingly these data points contribute significantly in the estimation of the function $f(x)$.

The optimization problem in Eq. 4.9 is then formulated as a quadratic programming problem by first determining the Lagrangian multiplier and using the Karush-Kuhn Tucker (KKT) conditions (Joachims 1999). Then the values for w and b can be estimated so that the linear function fit the training data can be identified. This instance of applying the constrained optimization framework is merely applicable for a linear kernel function and the constrained optimization framework differs for different kernel functions.

Similarly, a non-linear model can be used to satisfactorily model the data and this can be conducted by applying a non-linear function to relate the data into a high dimensional feature space where linear regression is conducted. Then the kernel method is applied to handle the problem of the curse of dimensionality and for non-linear problems the ε-insensitive loss function can be used to give (Gunn 1997):

$$\max_{\alpha,\alpha^*} W\left(\alpha,\alpha^*\right) = \max_{\alpha,\alpha^*} \sum_{i=1}^{l} \alpha^* \left(y_i - \varepsilon\right)$$

$$-\alpha_i \left(y_i + \varepsilon\right) - \frac{1}{2} \sum_{i=1}^{l} \sum_{j=1}^{l} \left(\alpha_i^* - \alpha_i\right) \left(\alpha_j^* - \alpha_j\right) K\left(x_i, x_j\right)$$

(4.12)

subject to

$$0 \leq \alpha_i, \alpha_i^* \leq C, i = 1, \ldots, l$$

$$\sum_{i=1}^{l} \left(\alpha_i - \alpha_i^*\right) = 0 \tag{4.13}$$

In Eqs. 4.12 and 4.13, K is the kernel function and α and α^* are Lagrangian multipliers. Solving these equations provides the Lagrangian multipliers and the regression equation can thus be written as follows (Gunn 1997):

$$f(x) = \sum_{SVs} \left(\bar{\alpha}_i - \bar{\alpha}_i^*\right) K\left(x_i, x\right) + \bar{b} \tag{4.14}$$

$$b = -\frac{1}{2} \sum_{i=1}^{l} \left(\alpha_i - \alpha_i^*\right) \left(K\left(x_i, x_r\right) + K\left(x_i, x_s\right)\right) \tag{4.15}$$

The least squares support vector toolbox was applied for the investigation (Suykens et al. 2002).

4.5 Applications of MLP, RBF and SVM to Economic Modeling

In Chap. 3, the Bayesian and the evidence frameworks were applied to build an automatic relevance determination (ARD) method. The ARD technique was then applied to identify the relevance of economic variables that are essential for driving the consumer price index. It was concluded that for the data analyzed using the MLP based ARD the variables driving the CPI are mining, transport, storage and communication, financial intermediation, insurance, real estate and business services, community, social and personal services, gross value added at basic prices, taxes less subsidies on products, affordability, economic growth, repo rate, gross domestic product, household consumption and investment.

Table 4.2 Characteristics of
the MLP network and the
results

Attributes	Number
Input nodes	12
Hidden nodes	8
Output nodes	1
Training time	3.45
Accuracy	84.6 %

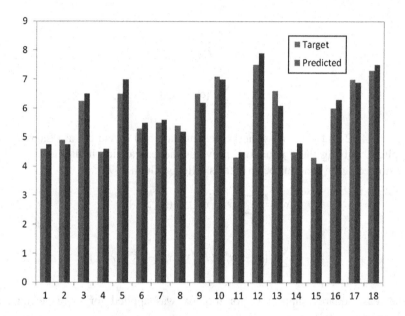

Fig. 4.3 Sample results achieved using the MLP network

Table 4.3 Characteristics of
the RBF network and the
results

Attributes	Number
Input nodes	12
Hidden nodes	8
Output nodes	1
Training time	0.36
Accuracy	82.7 %

In this chapter, these variables are used to predict the CPI using MLP, RBF, and
SVM. The architecture of the MLP neural network applied in this chapter is as
indicated in Table 4.2. The MLP had linear activation function in the output layer
and hyperbolic tangent function in the hidden nodes (Fig. 4.3). The architecture of
the RBF neural network, applied in this chapter, is as indicated in Table 4.3 and
Fig. 4.4. The RBF had a Gaussian basis function. The sample results obtained can
be viewed in Fig. 4.3.

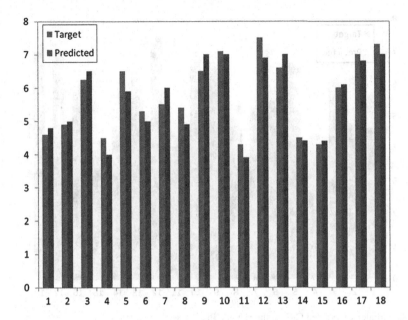

Fig. 4.4 Sample results achieved using the RBF network

Table 4.4 Different implementations of the SVM models

	C	e	Order	Kernel	Accuracy
Model 1	200	0.05	NA	Exponential RBF	89.39
Model 2	200	0.05	NA	Linear	87.81
Model 3	200	0.05	1	Spline	83.45
Model 4	200	0.05	2	RBF	85.73
Model 5	200	0.05	3	Polynomial	88.36

The architectures of the different SVM networks applied in this chapter are as indicated in Table 4.4. The C and e parameters were fixed at 200 and 0.05 respectively.

Figure 4.5 shows the sample results from the RBF based SVM.

The results indicate that for the data under consideration, the SVM models demonstrated better results followed by the MLP then the RBF. The SVM with the exponential RBF gave the best results.

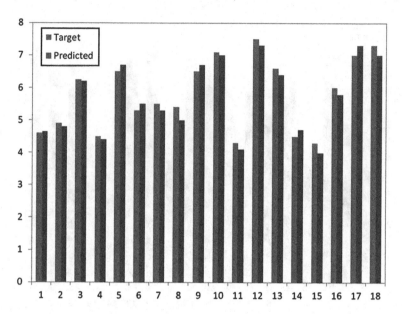

Fig. 4.5 Model 1 predicted CPI *vs.* the actual CPI

4.6 Conclusion

This chapter described the multi-layered perceptron, radial basis functions, and support vector machines and applied these to modeling the CPI. The results indicated that the SVM gave the best results followed by the MLP and then the RBF. The SVM with an exponential RBF gave the best results.

References

Ahn J-H, Rhee K-P, You Y-J (2012) A study on the collision avoidance of a ship using neural networks and fuzzy logic. Appl Ocean Res 37:162–173

Aïssa MBS, Musy O, Pereau J-C (2007) Modelling inflation persistence with periodicity changes in fixed and predetermined prices models. Econ Model 24:823–838

Anderson RG, Binner JM, Schmidt VA (2012) Connectionist-based rules describing the pass-through of individual goods prices into trend inflation in the United States. Econ Lett 117:174–177

Arebey M, Hannan MA, Begum RA, Basri H (2012) Solid waste bin level detection using gray level co-occurrence matrix feature extraction approach. J Environ Manag 104:9–18

Binner JM, Bissoondeeal RK, Elger T, Gazely AM, Mullineux AW (2005) A comparison of linear forecasting models and neural networks: an application to Euro inflation and Euro Divisia. Appl Econ 37:665–680

Bishop CM (1995) Neural networks for pattern recognition. Oxford University Press, Oxford

Buhmann MD, Ablowitz MJ (2003) Radial basis functions: theory and implementations. Cambridge University, Cambridge

Burnham KP, Anderson DR (2002) Model selection and multimodel inference: a practical information-theoretic approach, 2nd edn. Springer, New York

Cao Q, Ewing BT, Thompson MA (2012) Forecasting medical cost inflation rates: a model comparison approach. Decis Support Syst 53:154–160

Chang BR, Tsai HF (2008) Forecast approach using neural network adaptation to support vector regression grey model and generalized auto-regressive conditional heteroskedasticity. Expert Syst Appl 34:925–934

Chen C-S, Cheng M-Y, Wu Y-W (2012a) Seismic assessment of school buildings in Taiwan using the evolutionary support vector machine inference system. Expert Syst Appl 39:4102–4110

Chen Y, Xu J, Yang B, Zhao Y, He W (2012b) A novel method for prediction of protein interaction sites based on integrated RBF neural networks. Comput Biol Med 42:402–407

Chen Z-Y, Fan Z-P, Sun M (2012c) A hierarchical multiple kernel support vector machine for customer churn prediction using longitudinal behavioral data. Eur J Oper Res 223:461–472

Choudhary M, Haider A (2012) Neural network models for inflation forecasting: an appraisal. Appl Econ 44:2631–2635

Chuang C-C (2008) Extended support vector interval regression networks for interval input–output data. Inf Sci 178:871–891

De Gooijer JG, Vidiella-i-Anguera A (2003) Nonlinear stochastic inflation modelling using SEASETARs. Insur Math Econ 32:3–18

Dimirovski GM, Andreeski CJ (2006) How good ANN identification of post-stabilization inflation dynamics can be? In: Proceedings of the IEEE international conference on neural networks, Vancouver, pp 2098–2105

Drezet PML, Harrison RF (2001) A new method for sparsity control in support vector classification and regression. Pattern Recognit 34:111–125

Düzgün R (2010) Generalized regression neural networks for inflation forecasting. Int Res J Finance Econ 51:59–70

Ge L, Yin G (2012) Application of process neural network on consumer price index prediction. Adv Intell Soft Comput 137:427–432

Gidudu A, Hulley G, Marwala T (2007) Image classification using SVMs: one-against-one vs one-against-all. In: Proceedings of the 28th Asian conference on remote sensing, Kuala Lumpur, arXiv:0711.2914v1

Golbabai DA, Milev M (2012) Radial basis functions with application to finance: american put option under jump diffusion. Math Comput Model 55:1354–1362

Golmohammadi H, Dashtbozorgi Z, Acree WE Jr (2012) Quantitative structure–activity relationship prediction of blood-to-brain partitioning behavior using support vector machine. Eur J Pharm Sci 47:421–429

Golub GH, Van Loan CF (1996) Matrix computations. Johns Hopkins University Press, Baltimore

Gregoriou A, Kontonikas A (2009) Modeling the behaviour of inflation deviations from the target. Econ Model 26:90–95

Gryllias KC, Antoniadis IA (2012) A Support Vector Machine approach based on physical model training for rolling element bearing fault detection in industrial environments. Eng Appl Artif Intell 25:326–344

Gunn SR (1997) Support vector machines for classification and regression. Technical report, University of Southampton, Southampton

Guo G, Zhang JS (2007) Reducing examples to accelerate support vector regression. Pattern Recognit Lett 28:2173–2183

Guo X, Wang H, Yang F (2012) Thermal power financial environment risk forecast model by combined stock multi-indicators basis on RBF neural network. AASRI Procedia 1:519–524

Hartigan JA (1975) Clustering algorithms. Wiley, New York

Hartigan JA, Wong MA (1979) A K-means clustering algorithm. Appl Stat 28:100–108

Hassoun MH (1995) Fundamentals of artificial neural networks. MIT Press, Cambridge, MA

Haykin S (1999) Neural networks. Prentice-Hall, Upper Saddle River

Hsu C-F, Chiu C-J, Tsai J-Z (2012) Indirect adaptive self-organizing RBF neural controller design with a dynamical training approach. Expert Syst Appl 39:564–573

Hu T, Xiang D-H, Zhou D-X (2012) Online learning for quantile regression and support vector regression. J Stat Plan Inference 142:3107–3122

Huang C-F (2012a) A hybrid stock selection model using genetic algorithms and support vector regression. Appl Soft Comput J 12:807–818

Huang S (2012b) Based on genetic optimization of support vector machine consumer price index forecast. Adv Sci Lett 6:485–488

Ikbal S, Misra H, Hermansky H, Magimai-Doss M (2012) Phase AutoCorrelation (PAC) features for noise robust speech recognition. Speech Commun 54:867–880

Jing B, Ning-qiang J (2012) Power system's damping analysis based on RBF neural network1. Phys Procedia 24:1018–1023

Joachims J (1999) Making large-scale SVM learning practical. In: Scholkopf B, Burges CJC, Smola AJ (eds) Advances in kernel methods-support vector learning. MIT Press, Cambridge

Jochmann M, Koop G, Potter SM (2010) Modeling the dynamics of inflation compensation. J Empir Finance 17:157–167

Kapetanios G (2004) A note on modelling core inflation for the UK using a new dynamic factor estimation method and a large disaggregated price index dataset. Econ Lett 85:63–69

Karami A, Christianus A, Bahraminejad B, Gagné F, Courtenay SC (2012) Artificial neural network modeling of biomarkers to infer characteristics of contaminant exposure in *Clarias gariepinus*. Ecotoxicol Environ Saf 77:28–34

Khemchandani R, Jayadeva K, Chandra S (2009) Regularized least squares fuzzy support vector regression for financial time series forecasting. Expert Syst Appl 36:132–138

Kukal J, Van Quang T (2011) Modeling the CNB's monetary policy interest rate by artificial neural networks [Modelování Měnově Politické Úrokové Míry ČNB Neuronovými Sítěmi]. Politicka Ekonomie 59:810–829

Lang S (1987) Linear algebra. Springer, London

Li Z, Liu P, Wang W, Xu C (2012) Using support vector machine models for crash injury severity analysis. Accid Anal Prev 45:478–486

Lloyd SO (1982) Least squares quantization in PCM. IEEE Trans Inf Theory 28:129–137

Luo F, Huang S (2012) The application of the combinatorial and optimal networks on the prediction of the consumer price index. Adv Mater Res 488–489:886–891

Marcozzi MD, Choi S, Chen CS (2001) On the use of boundary conditions for variational formulations arising in financial mathematics. Appl Math Comput 124:197–214

Marwala T (2009) Computational intelligence for missing data imputation, estimation and management: knowledge optimization techniques. IGI Global Publications, New York

Marwala T (2010) Finite element model updating using computational intelligence techniques. Springer, London

Marwala T (2012) Condition monitoring using computational intelligence methods. Springer, London

Marwala T, Lagazio M (2011) Militarized conflict modeling using computational intelligence techniques. Springer, London

Mehrotra A, Peltonen T, Santos Rivera A (2010) Modelling inflation in china – a regional perspective. China Econ Rev 21:237–255

Mitra S (2012) Early warning prediction system for high inflation: an elitist neuro-genetic network model for the Indian economy. Neural Comput Appl. doi:10.1007/s00521-012-0895-4

Møller AF (1993) A scaled conjugate gradient algorithm for fast supervised learning. Neural Netw 6:525–533

Moore EH (1920) On the reciprocal of the general algebraic matrix. Bull Am Math Soc 26:394–395

Moshiri S, Cameron N (2000) Neural network versus econometric models in forecasting inflation. J Forecast 19:201–217

Nabney IT (2001) Netlab algorithms for pattern recognition. Springer, London

Oliveira ALI (2006) Estimation of software project effort with support vector regression. Neurocomputing 69:1749–1753

Orchel M (2012) Support vector regression based on data shifting. Neurocomputing 96:2–11

Penrose R (1955) A generalized inverse for matrices. Proc Camb Philos Soc 51:406–413

Pires MM, Marwala T (2004) Option pricing using neural networks and support vector machines. In: Proceedings of the IEEE International Conference on Systems, Man and Cybernetics, The Hague, 2004, pp 1279–1285

Price S, Nasim A (1999) Modelling inflation and the demand for money in Pakistan; cointegration and the causal structure. Econ Model 16:87–103

Rumpf T, Römer C, Weis M, Sökefeld M, Gerhards R, Plümer L (2012) Sequential support vector machine classification for small-grain weed species discrimination with special regard to cirsium arvense and galium aparine. Comput Electron Agric 80:89–96

Salem SB, Bacha K, Chaari A (2012) Support vector machine based decision for mechanical fault condition monitoring in induction motor using an advanced Hilbert-Park transform. ISA Trans 51:566–572

Santana JC (2006) Forecasting time series with neural networks: an application to the Colombian inflation [Predicción de Series Temporales con Redes Neuronales: Una Aplicación a la Inflación Colombiana]. Revista Colombiana de Estadistica 29:77–92

Singh A, Imtiyaz M, Isaac RK, Denis DM (2012) Comparison of soil and water assessment tool (SWAT) and multilayer perceptron (MLP) artificial neural network for predicting sediment yield in the Nagwa agricultural watershed in Jharkhand, India. Agric Water Manag 104:113–120

Souahlia S, Bacha K, Chaari A (2012) MLP neural network-based decision for power transformers fault diagnosis using an improved combination of Rogers and Doernenburg ratios DGA. Int J Electr Power Energy Syst 43:1346–1353

Suykens JAK, Van Gestel T, De Brabanter J, De Moor B, Vandewalle J (2002) Least squares support vector machines. World Scientific, Singapore

Trafalis TB, Ince H (2000) Support vector machine for regression and applications to financial forecasting. In: Proceedings of the IEEE International Joint Conference on Neural Networks, Como, 2000, pp 348–353

Tsai C-F, Wu J-W (2008) Using neural network ensembles for bankruptcy prediction and credit scoring. Expert Syst Appl 34:2639–2649

Tung WL, Quek C, Cheng P (2004) GenSo-EWS: a novel neural-fuzzy based early warning system for predicting bank failures. Neural Netw 17:567–587

Üstün B, Melssen WJ, Buydens LMC (2007) Visualisation and interpretation of Support Vector Regression models. Anal Chim Acta 595:299–309

Vapnik V (1995) The nature of statistical learning theory. Springer, Berlin

Vapnik V (1998) Statistical learning theory. Wiley-Interscience, New York

Wang H, Fan G (2010) Study on the model of CPI prediction based on BP neural network. In: Proceedings of the 2nd international conference on information technology and computer science, Kiev, pp 573–576

Wang Y, Wang B, Zhang X (2012) A new application of the support vector regression on the construction of financial conditions index to CPI prediction. Procedia Comput Sci 9:1263–1272

Welfe A (2000) Modeling inflation in Poland. Econ Model 17:375–385

Yang L, Zhang J (2012) Prediction study on anti-slide control of railway vehicle based on RBF neural networks. Physics Procedia 25:911–916

Yeganeh B, Motlagh MSP, Rashidi Y, Kamalan H (2012) Prediction of CO concentrations based on a hybrid Partial Least Square and Support Vector Machine model. Atmos Environ 55:357–365

Yu S, Wei Y-M, Wang K (2012) China's primary energy demands in 2020: predictions from an MPSO–RBF estimation model. Energy Convers Manag 61:59–66

Yümlü S, Gürgen FS, Okay N (2005) A comparison of global, recurrent and smoothed-piecewise neural models for Istanbul stock exchange (ISE) prediction. Pattern Recognit Lett 26:2093–2103

Zeng Z, Zhou J, Tao N, Feng L, Zhang C, Han X (2012) Support vector machines based defect recognition in SonicIR using 2D heat diffusion features. NDT E Int 47:116–123

Zhang J, Sato T, Iai S (2006) Support vector regression for on-line health monitoring of large-scale structures. Struct Saf 28:392–406

Zhong P (2012) Training robust support vector regression with smooth non-convex loss function. Optim Method Softw 27:1039–1058

Chapter 5
Bayesian Support Vector Machines for Economic Modeling: Application to Option Pricing

Abstract An option is the right, not the obligation, to buy or sell an underlying asset at a later date by fixing the price of the asset at the present moment. European styled options can be priced using the Black-Scholes equation and are only exercised at the end of the period but American options can be exercised at any time during the period and are, therefore, more complex due to the second random process they introduce. Support vector machines and multi-layered perceptron techniques are implemented using Bayesian technique to model American options and the results are compared.

5.1 Introduction

This chapter applies the Bayesian approach to model options. An option is a right but not an obligation to buy or sell an underlying asset at a later date but by fixing the price of that asset at the present moment. There are two types of options the American and European options and they differ with the specification on when they can be exercised. European option can only be exercised at the end of the period while American option can be exercised at any time during the life cycle of the options. Mathematically, European options can be exercised using what is known as the Black-Scholes equation which will be explained in the next section.

Lajbcygier and Connor (1997) applied neural networks and bootstrap methods to estimate the difference between the standard option-pricing model and intra-day option prices for stock index option futures. Confidence intervals resulting from the use of bootstrap approaches were applied for trading.

Anders et al. (1998) applied neural networks to improve the pricing of options on the German stock index. The results demonstrated better out-of-sample performance than the Black-Scholes approach.

Ghaziri et al. (2000) applied a multi-layer neural network and neuro-fuzzy network to price S&P 500 index call options. They compared the results with

T. Marwala, *Economic Modeling Using Artificial Intelligence Methods*, Advanced
Information and Knowledge Processing, DOI 10.1007/978-1-4471-5010-7_5,
© Springer-Verlag London 2013

the Black-Scholes approach. They observed that neural network method performed better than the Black-Scholes model as long as adequate amount of patterns were in the training data.

Liang et al. (2006) applied a three-layer feed-forward neural network to improve the option pricing performance. When applied to model the Hong Kong option market, the model gave better results than the conventional option pricing models.

Dindar and Marwala (2004) applied multi-layer perceptron (MLP) and radial basis functions (RBF) to model options. The MLP and RBF architectures were optimized using particle swarm optimization. The results showed that optimized RBF and MLP networks achieved better results than both the un-optimized networks and the committee of networks.

Pires and Marwala (2005) applied Bayesian multi-layer perceptrons and Bayesian support vector machines (SVM) for pricing American option. Their work is the basis of this chapter. Furthermore, Pires and Marwala (2004) applied the standard MLP and SVM for pricing American option pricing. Kohler et al. (2010) applied neural networks for pricing high-dimensional American options. They assumed that the price of the underlying asset were driven by Markov processes and applied the Monte Carlo method to produce simulated data of these price processes. They then applied standard neural networks to approximate option prices from this data and the results were found to be good.

Nqi and Lidouh (2010) successful applied neural networks to price option with monotonicity and convexity constraints, while Gradojevic et al. (2009) successfully applied neural networks for option pricing with modular neural networks.

Andreou et al. (2008) applied artificial neural networks to price European options and compared these to the Black-Scholes method. The results indicated that there existed profitable opportunities even in the presence of transaction costs. Thomaidis et al. (2007) compared neural network model selection strategies for pricing the S&P 500 stock index options.

Amornwattana et al. (2007) applied a hybrid option pricing model that used neural network for improving the estimation of option market prices. Neural network approximated volatility and another neural network was used to value the difference between the Black-Scholes model and the actual market option prices. The proposed method performed better than the Black-Scholes approach with historical volatility or the Black-Scholes approach model with volatility estimated by the neural network.

Zhang and Lin (2007) applied wavelet neural network to price European options. They compared this approach to MLP models and the Black-Scholes model on pricing the Hang Seng index call options. The results indicated that wavelet neural network was better than the MLP and the Black-Scholes model.

Other successful applications of neural networks for option pricing include using neural networks and multinomial tree (Chen and Magdon-Ismail 2006), neural networks and parametric models (Panayiotis et al. 2004), regularized neural networks (Choi et al. 2004), and for improving option pricing with product constrained hybrid neural networks (Lajbcygier 2004).

Shen et al. (2012) used support vector machine for option pricing and compared this to the Black-Scholes equation. The results indicated high accuracy, robustness, stability, and the ability to deal with non-linearity and non-stationarity characteristics. Li (2011) successfully applied SVM to combine various estimating methods to approximate option price, thereby, avoiding weaknesses of each method.

Liang et al. (2009) applied neural networks and SVM to improve option pricing. They proposed a model based on neural networks and support vector regressions. They implemented this method on the Hong Kong option market and the results showed improved forecasting accuracy. Wang and Huang (2006) successfully applied a hybrid wavelet-SVM for modeling derivatives valuation.

5.2 Black-Scholes Model

As described by Pires (2005), an option is a financial derivative of a different financial security such as stock, credit, interest rate or exchange rate. Any of these instruments are the primary asset of an option (Ross et al. 2001). Options are called derivatives for the reason that they are derived from other financial securities (Hull 2003). An option provides for the right, but not an obligation to the owner of the option, to buy or sell the asset at a later time called the maturity date by entering into a contract that specifies a price for the underlying asset now. This agreed price is known as the strike price. Because of their high value, options are treasured and organizations pay a premium known as the price of the option to possess them. There are two types of options and these are call and put options. A call option is when a person intends to buy the underlying asset, while a put option is when the individual intends to sell the underlying asset (Hull 2003).

Options are utilized daily by organizations to hedge their financial risk. For example, consider a firm with an exposure to foreign trade. This company's financial risk is governed by the exchange rate and, if the rate changes dramatically, the company may not be able to meet its financial obligations. A firm can normally use an option to protect itself by purchasing a call option and thus giving the firm the option of a fixed exchange rate.

This practice is called hedging and makes options very treasured and, therefore, organizations pay a premium called the price of the option (Hull 2003). Other financial mechanisms that can be used for hedging include forwards and futures (Ross et al. 2001). There are two types of options and these are European and American options (Hull 2003; Pires 2005). European options only permit the option owner to exercise them on the expiry date of the contract, while American options permit the owners to exercise the option on any date between accepting and the end date of the contract (Hull 2003). American options are more valuable and introduce a second random process into the model because of their flexibility on the date they can be exercised (Jarrow and Turnbull 2000).

Black, Scholes and Merton presented the Black-Scholes model for option pricing (Black and Scholes 1973; Merton 1973) and proposed the option pricing formula for European options. The Black-Scholes model is premised on the following assumptions (Black and Scholes 1973; Merton 1973):

- Absence of arbitrage.
- Cash can be borrowed and lent at a known constant risk-free interest rate.
- Stock can be bought and sold.
- Transactions have fees or costs.
- The stock price is described by a geometric Brownian motion with constant drift and volatility.
- The underlying security does declare dividend.

The Black-Scholes model offers a mechanism of approximating the underlying asset's volatility which can be predicted using the Black-Scholes equation. The model offered the first options pricing formula and influenced the manner in which traders price and hedge options. American options are used extensively by traders and many traders use the Black-Scholes model together with some sampling process such as Monte Carlo simulation, which will be described later in the chapter.

Chen et al. (2012) proposed a spectral element technique for solving partial integro-differential equations for pricing European options under the Black-Scholes and Merton jump diffusion models. The proposed technique was found to be flexible for treating different boundary conditions and non-smooth initial conditions and was found to be efficient and accurate.

Jeunesse and Jourdain (2012) studied the regularity of the value function for the American Put option when the underlying asset declares a discrete dividend at a prescribed date, while Iazzolino and Fortino (2012) proposed a correction to the Black-Scholes model to assess credit risk. Thapa et al. (2012) studied parameters in the Black-Scholes option pricing model and established the existence, uniqueness, and continuous dependence of the weak solution of the Black-Scholes model.

Fatone et al. (2012) applied statistical tests to calibrate the Black-Scholes asset dynamics model to price options with uncertain volatility, while Yousuf-Khaliq and Kleefeld (2012) applied numerical approximation of nonlinear Black-Scholes model for exotic path-dependent American options with transaction cost.

The Black-Scholes equation can be expressed mathematically as follows (Hull 2003; Black and Scholes 1973):

$$\frac{\partial V}{\partial t} + \frac{1}{2}S^2\sigma^2\frac{\partial^2 V}{\partial S^2} + rS\frac{\partial V}{\partial S} - rV = 0 \tag{5.1}$$

Here, S is the price of the stock, V is the price of the derivative as a function of time and the stock price σ is the vitality of the stock return, and r is the annualized risk-free interest rate. Equation 5.1 can be solved for both call and put options. The price for a call option for a non-dividend declaring underlying stock can be mathematically represented as follows (Hull 2003):

$$C(S,t) = N(d_1)S - N(d_2)Ke^{-r(T-t)} \tag{5.2}$$

here,

$$d_1 = \frac{\ln\left(\frac{S}{K}\right) + \left(r + \frac{\sigma^2}{2}\right)(T-t)}{\sigma\sqrt{T-t}} \tag{5.3}$$

and

$$d_2 = \frac{\ln\left(\frac{S}{K}\right) + \left(r - \frac{\sigma^2}{2}\right)(T-t)}{\sigma\sqrt{T-t}} \tag{5.4}$$

The price of an equivalent put option can be written mathematically as follows (Hull 2003):

$$
\begin{aligned}
P(S,t) &= Ke^{-r(T-t)} - S + C(S,t) \\
&= N(-d_2)Ke^{-r(T-t)} - N(-d_1)S
\end{aligned}
\tag{5.5}
$$

where

$$N(x) = \frac{1}{\sqrt{2\pi}} \int_{-\infty}^{x} e^{-\frac{z^2}{2}} dz \tag{5.6}$$

and $T-t$ is the time to maturity, and K is the strike price. In this chapter, we implement Bayesian neural networks and Bayesian support machines for American option pricing.

5.3 Bayesian Neural Networks

In this chapter, a multi-layer perceptron neural network model is used for option pricing. The MLP is formulated in the Bayesian framework and trained using the hybrid Monte Carlo technique (Marwala 2001, 2007, 2009, 2010, 2012; Marwala and Lagazio 2011).

Venegas-Martínez (2005) proposed a Bayesian model for pricing options with prior volatility information. A number of estimated formulas for valuing European call options on the basis of asymptotic and polynomial estimates of Bessel functions were presented. Flynn et al. (2005) presented a Bayesian framework for pricing Australian S&P 200 options. The results showed that time-varying volatility, leptokurtosis, and a small degree of negative skewness were priced in Australian stock market options.

Bauwens and Lubrano (2002) proposed Bayesian option pricing models using asymmetric GARCH models, while Foster and Whiteman (1999) applied Bayesian option pricing models to the soybean market.

In this chapter, a multi-layer perceptron is applied to map the spot price, strike price, the risk-free rate of interest, the time to maturity of the contract (the time difference between when the contract is taken and when the contract expires) and the volatility of the underlying asset (x) and the price of the call option (y). This relationship between the y, and x, may be expressed as follows (Bishop 1995; Marwala 2009, 2012; Marwala and Lagazio 2011):

$$y = f_{outer} \left(\sum_{j=1}^{M} w_{kj}^{(2)} f_{inner} \left(\sum_{i=1}^{d} w_{ji}^{(1)} x_i + w_{j0}^{(1)} \right) + w_{k0}^{(2)} \right) \tag{5.7}$$

Here, $w_{ji}^{(1)}$ and $w_{kj}^{(2)}$ designate the weights in the first and second layers, correspondingly, moving from input i to hidden unit j, M is the number of hidden units, d is the number of output units, whereas $w_{j0}^{(1)}$ designates the bias for the hidden unit j and $w_{k0}^{(2)}$ designates the bias for the output unit k.

Selecting suitable network design is an imperative requirement for model building. When selecting a correct MLP model, an essential decision lies in the choice of the correct number of hidden units (M) and the activation functions. A large value of M gives flexible networks, which learn the noise in the data. On the contrary, a small M gives networks that are incapable of modeling complex relationships. Identifying the weights and biases in the MLP neural networks may be framed in the Bayesian context as (MacKay 1991, 1992; Bishop 1995; Lagazio and Marwala 2005; Marwala 2009, 2012; Marwala and Lagazio 2011):

$$P(w|[D]) = \frac{P([D]|w)P(w)}{P([D])} \tag{5.8}$$

$P(w)$ is the probability distribution function of the weight-space before the data is factored into account and it is also the *prior distribution function* and $[D] \equiv (y_1, \ldots, y_N)$ is a matrix comprising the option prices. $P(w|[D])$ is the posterior probability distribution function after the data have been seen, $P([D]|w)$ is the likelihood function and $P([D])$ is the normalization function, also called the "evidence". For the MLP, Eq. 5.8 may be expanded by applying the sum-of squares of the error function to give (MacKay 1992; Bishop 1995; Marwala 2009):

$$P(w|[D]) = \frac{1}{Z_s} \exp \left(\beta \sum_{n}^{N} \sum_{k}^{K} \{t_{nk} - y_{nk}\}^2 - \sum_{j}^{W} \frac{\alpha_j}{2} w_j^2 \right) \tag{5.9}$$

where

$$Z_S(\alpha, \beta) = \left(\frac{2\pi}{\beta} \right)^{\frac{N}{2}} + \left(\frac{2\pi}{\alpha} \right)^{\frac{W}{2}} \tag{5.10}$$

The sum-of-squares of the error function is applied because of its regression advantages. A weight-decay is assumed for the prior distribution as it penalizes the weights with large magnitudes. In Eq. 5.9, n is the index for the training example, hyper-parameter β is the data contribution to the error, k is the index for the output units, t_{nk} is the target output corresponding to the nth training example and kth output unit and y is the forecasted option price. The parameter α_j is a hyper-parameter, which influences the contribution of the regularization term to the training error. Equation 5.9 may be solved by sampling the posterior probability space using methods like the Monte Carlo technique, simulated annealing, the genetic Monte Carlo method, or the hybrid Monte Carlo method (Marwala 2010, 2012). The next section describes the hybrid Monte Carlo method which is used in this chapter.

5.4 Hybrid Monte Carlo (HMC)

This chapter applies the HMC method to approximate the posterior probability of the weight vectors, given the observed data as indicated in Eq. 5.9. The HMC technique uses the gradient of the mean-square error that is estimated using a back-propagation method. Using the gradient method assures that the simulation samples through the areas of higher probabilities and, therefore, accelerates convergence to a stationary probability distribution function. This technique is a Markov chain method. The HMC consists of dynamic and stochastic moves. Stochastic moves allow the method to sample states with different total energy, while the dynamic moves apply the Hamiltonian dynamics and allow sampling states with constant total energy. The HMC method can be considered as a hybrid of Monte Carlo method and gradient search.

Misev et al. (2012) applied HMC for modeling condensed phases in high performance computing environment, while Wang et al. (2012) applied HMC and support vector machines for face detection demonstrating the reduction of the training time. Leermakers et al. (2012) applied the HMC to model a thin layer of a polyelectrolyte gel near an adsorbing surface, while Zhang et al. (2012) applied the HMC to estimate geotechnical model uncertainty, and (Niu et al. 2012) applied the HMC technique in short-term load forecasting.

In *Mechanics* the positions and the momentum of all molecules at a particular time in a physical system is called the *state space* of the system. The positions of the molecules define the potential energy of the system and the momentum articulates the kinetic energy of the system. What is called the *canonical distribution* of the 'potential energy' in statistical mechanics is the posterior distribution in this chapter. The canonical distribution of the system's kinetic energy is (Neal 1993; Bishop 1995; Marwala 2009, 2012; Marwala and Lagazio 2011):

$$P\left(\{p\}\right) = \frac{1}{Z_K} \exp\left(-K\left(\{p\}\right)\right)$$

$$= (2\pi)^{-n/2} \exp\left(-\frac{1}{2}\sum_i p_i^2\right) \tag{5.11}$$

Here, p_i is the momentum of the ith variable and p should not to be confused with P, which indicates the probability. In this chapter, p_i is a fictional parameter that presents the technique in a molecular dynamics context. The weight vector $\{w\}$ and momentum vector $\{p\}$ are of the same dimension. The sum of the kinetic and potential energy is known as the *Hamiltonian* of the system and can be mathematically designated as follows (Neal 1993; Bishop 1995; Marwala 2009, 2012; Marwala and Lagazio 2011):

$$H(w, p) = \beta \sum_{k}^{N} \sum^{K} \{y_{nk} - t_{nk}\}^2 + \frac{\alpha}{2} \sum_{j=1}^{W} w_j^2 + \frac{1}{2} \sum_{i}^{W} p_i^2 \qquad (5.12)$$

Here, the first two terms are the potential energy of the system and the last term is the kinetic energy. The canonical distribution over the phase space, *i.e.*, position and momentum, can be written as follows (Neal 1993; Bishop 1995; Marwala 2009, 2012; Marwala and Lagazio 2011):

$$P(w, p) = \frac{1}{Z} \exp(-H(w, p)) = P(w|D) P(p) \qquad (5.13)$$

The dynamics in the phase space may be expressed in the Hamiltonian dynamics by deriving the derivative of the 'position' and 'momentum' in terms of fictional time τ. The term 'position' used here is the network weights. The dynamics of the system accordingly can be written by using Hamiltonian dynamics as follows (Neal 1993; Bishop 1995; Marwala 2009, 2012; Marwala and Lagazio 2011):

$$\frac{dw_i}{d\tau} = +\frac{\partial H}{\partial p_i} = p_i \qquad (5.14)$$

$$\frac{dp_i}{d\tau} = +\frac{\partial H}{\partial w_i} = -\frac{\partial E}{\partial p_i} \qquad (5.15)$$

The dynamics, stated in Eqs. 5.14 and 5.15, cannot be realized precisely. These equations are discretized by applying a 'leapfrog' method. The leapfrog discretization of Eqs. 5.14 and 5.15 may be written as follows (Neal 1993; Bishop 1995; Marwala 2009, 2012; Marwala and Lagazio 2011):

$$\hat{p}_i \left(\tau + \frac{\varepsilon}{2} \right) = \hat{p}_i(\tau) - \frac{\varepsilon}{2} \frac{\partial E}{\partial w_i} (\hat{w}(\tau)) \qquad (5.16)$$

$$\hat{w}_i(\tau + \varepsilon) = \hat{w}_i(\tau) + \varepsilon \hat{p}_i \left(\tau + \frac{\varepsilon}{2} \right) \qquad (5.17)$$

$$\hat{p}_i(\tau + \varepsilon) = \hat{p}_i \left(\tau + \frac{\varepsilon}{2} \right) - \frac{\varepsilon}{2} \frac{\partial E}{\partial w_i} (\hat{w}(\tau + \varepsilon)) \qquad (5.18)$$

By using Eq. 5.16, the leapfrog moves a slight half step for the momentum vector, $\{p\}$, and, using Eq. 5.17, moves a full step for the 'position', $\{w\}$, and, by using Eq. 5.18, moves a half step for the momentum vector, $\{p\}$. The amalgamation of these three steps give a single leapfrog iteration that estimates the *'position'* and *'momentum'* of a system at time $\tau + \varepsilon$ from the network weight vector and 'momentum' at time τ. This discretization process is reversible in time, it nearly conserves the Hamiltonian, and conserves the volume in the phase space, as required by Liouville's theorem (Neal 1993).

The HMC method functions by following a sequence of routes from an initial state, *i.e.*, *'positions'* and *'momentum'*, and moving in some direction in the state space for a given length of time and accepting the final state by applying the Metropolis algorithm which is written as follows (Metropolis et al. 1953; Marwala 2009, 2010, 2012; Marwala and Lagazio 2011):

$$if \ E_{new} < E_{old} \ accept \ state \ (s_{new})$$

$$else$$

$$accept \ (s_{new}) \ with \ probability$$

$$\exp\{-(E_{new} - E_{old})\} \tag{5.19}$$

In Eq. 5.19, states with high probability form the bulk of the Markov chain and those with low probability form the smaller part of the Markov chain. The validity of the hybrid Monte Carlo depends on three properties of the Hamiltonian dynamics which were described by Neal (1993), Bishop (1995), Marwala (2009), Marwala and Lagazio (2011), and Marwala (2012) as follows:

1. Time reversibility: it is invariant under $t \rightarrow -t, p \rightarrow -p$.
2. Conservation of energy: the $H(w, p)$ is the same at all times.
3. Conservation of state space volumes due to Liouville's theorem (Neal 1993).

For a specified leapfrog step size, ε_0, and the number of leapfrog steps, L, the dynamic transition of the hybrid Monte Carlo method is implemented as described by Neal (1993), Bishop (1995), Marwala (2009), Marwala and Lagazio (2011), as well as Marwala (2012):

- Select randomly the direction of the trajectory, λ, to be either -1 for a backwards trajectory or $+1$ for forwards trajectory.
- Beginning from the first state, $(\{w\}, \{p\})$, execute L leapfrog steps with the step size $\varepsilon = \varepsilon_0(1 + 0.1k)$ leading to state $(\{w\}^*, \{p\}^*)$. Here, ε_0 is a chosen fixed step size and k is a number chosen from a uniform distribution and is between 0 and 1.
- Reject or accept $(\{w\}^*, \{p\}^*)$ by using the Metropolis criterion. If the state is accepted then the new state becomes $(\{w\}^*, \{p\}^*)$. If rejected, the old state, $(\{w\}, \{p\})$, is reserved as the new state.

The HMC technique uses the gradient information in Step 2 and the advantages of using this gradient is so that the HMC trajectories move in the direction of high probabilities, resulting in an improved acceptance rate of samples (Bishop 1995).

The number of leapfrog steps, L, must be higher than one to permit a fast exploration of the state space. The choice of ε_0 and L impacts on the speed at which the simulation converges to a stationary distribution and the correlation between the states accepted. The leapfrog discretization doesn't present systematic errors because of infrequent rejection of states that end with the increase of the Hamiltonian. In the HMC technique, the step size $\varepsilon = \varepsilon_0(1 + 0.1k)$ where k is uniformly distributed between 0 and 1 is not fixed and this guarantees that the step size for each trajectory is altered with the aim of accepting states that have no high correlation. The same result can be realized by varying the leapfrog steps but in this chapter only the step size is altered. The use of the Bayesian method in neural networks leads to weight vectors that have a particular mean and standard deviation. Accordingly, the output parameters have a probability distribution. Resulting from the rules of probability theory, the distribution of the output vector $\{y\}$ for a given input vector $\{x\}$ can be expressed in the following form as explained in Bishop (1995), Marwala (2009), Marwala and Lagazio (2011), and Marwala (2012):

$$p(\{y\}|\{x\}, D) = \int p(\{y\}|\{x\}, \{w\}) p(\{w\}|D) d\{w\} \qquad (5.20)$$

The HMC technique was employed to identify the distribution of the weight vectors, and subsequently, of the output parameters. The integral in Eq. 5.20 may be approximated as follows (Bishop 1995; Neal 1993; Marwala 2009, 2012; Marwala and Lagazio 2011):

$$I \equiv \frac{1}{L} \sum_{i=1}^{L} f\left(\{w\}_i\right) \qquad (5.21)$$

Here, L is the number of retained states and f is the MLP network. The use of a Bayesian approach to the neural network leads to the mapping of the weight vector between the input and output having a probability distribution.

5.5   Bayesian Support Vector Machines

In support vector regression, as it was described in Chap. 4, we seek to identify small values for w with slack variables ξ_i, ξ_i^* in order to ensure that particular infeasible constraints in the minimization of the Euclidean norm can be applied. This problem can be formulated as follows (Schölkopf et al. 1999; Schölkopf and Smola 2002):

$$\min \frac{1}{2} \|w\|^2 + C \sum_{i=1}^{l} (\xi_i + \xi_i{}^*) \tag{5.22}$$

$$subject\ to \begin{cases} y_i - \langle w, x_i \rangle - b \leq & \varepsilon + \xi_i \\ \langle w, x_i \rangle + b - y_i \leq & \varepsilon + \xi_i{}^* \\ \xi_i, \xi_i{}^* \geq & 0 \end{cases} \tag{5.23}$$

where l is the number of training points used. The constraints in Eq. 5.23 are associated with an ε-insensitive loss function used to penalize specific training data that are outside of the bound given by ε which is a value chosen. The ε-insensitive loss function which is given by (Gunn 1997; Shawe-Taylor and Cristianini 2004):

$$|\xi|_\varepsilon = \begin{cases} 0 & if \quad |\xi| \leq \varepsilon \\ |\xi| - \varepsilon & otherwise \end{cases} \tag{5.24}$$

As it was described in Chap. 4, in support vector machines we identify a number of functions $f(x)$ to relate training inputs to training outputs and these functions are called kernel functions. In this chapter, we implement the radial basis function kernel which is written as follows (Gunn 1997; Steinwart and Christmann 2008):

$$k\left(x_i, x_j\right) = \exp\left(-\gamma \|x_i - x_j\|^2\right) \ for\ \gamma > 0 \tag{5.25}$$

A non-linear model can be applied to model the data and this can be achieved by using a non-linear function to relate the data into a high dimensional feature space where linear regression is conducted. Then the kernel method is applied to handle the problem of the curse of dimensionality and for non-linear problems the ε-insensitive loss function can be used to give the following optimization problem (Gunn 1997):

$$\max_{\alpha, \alpha^*} W\left(\alpha, \alpha^*\right) = \max_{\alpha, \alpha^*} \sum_{i=1}^{l} \alpha^* \left(y_i - \varepsilon\right) - \alpha_i \left(y_i + \varepsilon\right)$$

$$-\frac{1}{2} \sum_{i=1}^{l} \sum_{j=1}^{l} (\alpha_i^* - \alpha_i)(\alpha_j^* - \alpha_j) K(x_i, x_j) \tag{5.26}$$

subject to

$$0 \leq \alpha_i, \alpha_i^* \leq C,\ i = 1, \ldots, l$$

$$\sum_{i=1}^{l} (\alpha_i - \alpha_i^*) = 0 \tag{5.27}$$

Here, α_i and α_i^* are Lagrange's multipliers and C is the capacity. The regression equation can thus be written as follows (Gunn 1997):

$$f(x) = \sum_{i=1}^{l} \left(\alpha_i - \alpha_i^* \right) K \left(x_i, x_j \right) + b \tag{5.28}$$

where

$$b = -\frac{1}{2} \sum_{i=1}^{l} \left(\alpha_i - \alpha_i^* \right) \left(K \left(x_i, x_r \right) + K \left(x_i, x_s \right) \right) \tag{5.29}$$

5.5.1 Monte Carlo Method

In this chapter, we sample through the beta values space using Markov Chain Monte Carlo (MCMC) simulation to identify the distribution of the beta values. This is done to estimate Eq. 5.8 where the w is the vector containing the beta values. Before we describe the MCMC, it is important to describe the Monte Carlo method. Monte Carlo methods are applied to simulate complex systems and form a class of numerical techniques that depends on recurrent random sampling to estimate the results. Because they depend on recurrent estimation of random numbers, these techniques are appropriate for estimating the results using computers and are applied when it is mathematically infeasible to estimate a solution using a closed form solution (Marwala and Lagazio 2011; Marwala 2012).

Caporin et al. (2012) applied Monte Carlo method to price energy and temperature Quanto options. Quanto options allow for the correlation between energy consumption and weather conditions and, thereby, allow price and weather risk to be managed. They proposed a Monte Carlo pricing model which relaxes some of the assumptions made by the Black-Scholes model. Hu and Chen (2012) successfully applied Quasi-Monte Carlo simulation for Asian basket option pricing, while Caramellino and Zanette (2011) successfully applied a Monte Carlo approach for pricing and hedging American options in high dimensions. Holčapek and Tichý (2011) applied the Monte Carlo technique for option pricing with fuzzy parameters. Other applications of the Monte Carlo method to option pricing include pricing of Asian options using a Trapezium scheme (Seghiouer et al. 2011) and pricing options using sequential Monte Carlo (Jasra and del Moral 2011).

Monte Carlo simulation methods are beneficial in analyzing systems with a large number of degrees of freedom and uncertain inputs in disciplines such as mechanical engineering, electrical engineering, and computer science (Robert and Casella 2004). The Monte Carlo method usually follows the following steps (Robert and Casella 2004; Marwala and Lagazio 2011; Marwala 2012):

- The input space should be defined.
- Randomly produce inputs from the input space by using a chosen probability distribution.
- Apply the input for the deterministic computing.
- Integrate the results of the individual computation to estimate the final result.

5.5.2 Markov Chain Monte Carlo Method

A Monte Carlo method which is used in this chapter to sample the probability distribution of the beta factor in the SVM in this chapter is the Markov Chain Monte Carlo (MCMC) method. The MCMC is a random walk Monte Carlo method which produces a Markov chain to identify an equilibrium distribution. The MCMC encompasses a Markov process and a Monte Carlo simulation (Liesenfeld and Richard 2008).

Verrall and Wüthrich (2012) applied a reversible jump MCMC technique to reduce parameters for setting claims reserves for the outstanding loss liabilities. The proposed technique was observed to describe parameter reduction and tail factor approximation in the claims reserving procedure and predicted distribution of the outstanding loss liabilities. Kojima and Usami (2012) applied the Markov Chain Monte Carlo technique for solving the electromagnetic interrogation problem, while Fishman (2012) applied the MCMC for counting contingency tables, and Hettiarachchi et al. (2012) applied the MCMC approach to analyze of EEG data.

The MCMC method is applied by considering a system whose evolution is described by a stochastic process involving random variables $\{w_1, w_2, w_3, \ldots, w_i\}$ where a random variable w_i occupies a state w at discrete time i. The total number of all possible states that all random variables can inhabit is named a *state space*. If the probability that the system is in state w_{i+1} at time $i+1$ depends entirely on the fact that it was in state w_i at time i, then the random variables $\{w_1, w_2, w_3, \ldots, w_i\}$ form a Markov chain. The transition between states is realized by adding random noise (ε) to the current state as follows (Bishop 1995; Marwala 2010, 2012; Marwala and Lagazio 2011):

$$w_{i+1} = w_i + \varepsilon \tag{5.30}$$

When the current state has been reached, it is either accepted or rejected using the Metropolis algorithm which was described in Eq. 5.19.

5.6 Experimental Investigation

The data used were obtained from the South African Futures Exchange (Anonymous 2004). This results are based on the work conducted Pires (2005) and Pires and Marwala (2004, 2005). The data that was analyzed was for all call options for the

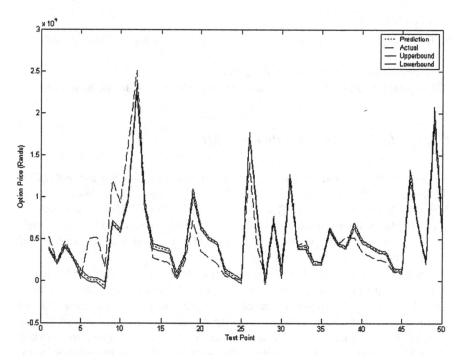

Fig. 5.1 Results obtained from the Bayesian MLP

period of January 2001 to December 2003. In this chapter, Bayesian multi-layered perceptrons and Bayesian SVM were used to map the relationship between the spot price, strike price, the risk-free rate of interest, time to maturity of the contract (the time difference between when the contract is taken and when the contract expires) and the volatility of the underlying asset to the call price. The MLP network which was used had seven inputs, 40 hidden nodes, and one output. The hidden nodes were hyperbolic tangent function while the output activation function was linear. In total, 300 samples were used for training and 300 were used for testing the methods. The weight decay value was 0.05 number of hybrid Monte Carlo samples was 1,000, step size of 0.0005 and number of omitted samples at the start of the hybrid Monte Carlo simulation was 500. The results obtained are indicated in Fig. 5.1 (Pires and Marwala 2005).

The SVM was implemented and the following parameters were chosen: Kernel: radial basis function, number of training points: 400, Capacity: 10, and ε-insensitivity of 0.005. The results obtained are indicated in Fig. 5.2 (Pires and Marwala 2005). The number of samples which is under 5 % was 42 for the SVM, while it was 56 for the MLP. The number of samples which was under 10 % is 82 for the SVM, while it was 112 for the MLP. These results indicate that the MLP performs better than the SVM.

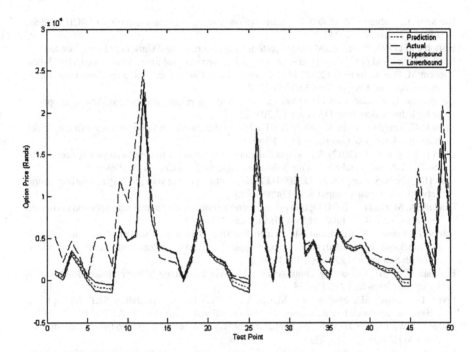

Fig. 5.2 Results obtained from the Bayesian SVM

5.7   Conclusion

As described, an option is the right, not the obligation, to buy or sell an underlying asset at a later date but by fixing the price of the asset at the present moment. Support vector machines and the multi-layered perceptron techniques were implemented using the Bayesian technique to model American options and the results were compared. The results obtained suggested that the MLP gave better results than the SVM.

References

Amornwattana S, Enke D, Dagli CH (2007) A hybrid option pricing model using a neural network for estimating volatility. Int J Gen Syst 36:558–573

Anders U, Korn O, Schmitt C (1998) Improving the pricing of options: a neural network approach. J Forecast 17:369–388

Andreou PC, Charalambous C, Martzoukos SH (2008) Pricing and trading European options by combining artificial neural networks and parametric models with implied parameters. Eur J Oper Res 185:1415–1433

Anonymous (2004) South African futures exchange. http://www.safex.co.za/. Last accessed Oct 2012

Bauwens L, Lubrano M (2002) Bayesian option pricing using asymmetric GARCH models. J Empir Financ 9:321–342

Bishop CM (1995) Neural networks for pattern recognition. Oxford University Press, London

Black F, Scholes M (1973) The pricing of options and corporate liabilities. J Polit Econ 81:637–654

Caporin M, Preś J, Torro H (2012) Model based Monte Carlo pricing of energy and temperature quanto options. Energy Econ 34:1700–1712

Caramellino L, Zanette A (2011) Monte Carlo methods for pricing and hedging American options in high dimension. Risk Decis Anal 2:207–220

Chen H-C, Magdon-Ismail M (2006) NN-OPT: neural network for option pricing using multinomial tree. Lect Note Comput Sci 4234:360–369

Chen F, Shen J, Yu H (2012) A new spectral element method for pricing European options under the Black-Scholes and Merton jump diffusion models. J Sci Comput 52:499–518

Choi H-J, Lee H-S, Han G-S, Lee J (2004) Efficient option pricing via a globally regularized neural network. Lect Note Comput Sci 3174:988–993

Dindar ZA, Marwala T (2004) Option pricing using a committee of neural networks and optimized networks. Proc IEEE Int Conf Syst Man Cybern 1:434–438

Fatone L, Mariani F, Recchioni MC, Zirilli F (2012) The use of statistical tests to calibrate the Black-Scholes asset dynamics model applied to pricing options with uncertain volatility. J Probab Stat art. no. 931609, 2012:1–20

Fishman GS (2012) Counting contingency tables via multistage Markov chain Monte Carlo. J Comput Graph Stat 21:713–738

Flynn DB, Grose SD, Martin GM, Martin VL (2005) Pricing Australian S&P 200 options: a Bayesian approach based on generalized distributional forms. Aust N Z J Stat 47:101–117

Foster FD, Whiteman CH (1999) An application of Bayesian option pricing to the soybean market. Am J Agric Econ 81:722–727

Ghaziri H, Elfakhani S, Assi J (2000) Neural networks approach to pricing options. Neural Netw World 10:271–277

Gradojevic N, Gençay R, Kukolj D (2009) Option pricing with modular neural networks. IEEE Trans Neural Netw 20:626–637

Gunn SR (1997) Support vector machines for classification and regression. Technical report, University of Southampton, Southampton

Hettiarachchi I, Mohamed S, Nahavandi S (2012) A marginalised Markov Chain Monte Carlo approach for model based analysis of EEG data. In: Proceedings of the international symposium on biomedical imaging, Barcelona, 2012, pp 1539–1542

Holčapek M, Tichý T (2011) Option pricing with fuzzy parameters via Monte Carlo simulation. Commun Comput Inf Sci 211:25–33

Hu Y-H, Chen D-Q (2012) Parallel randomized Quasi-Monte Carlo simulation for Asian basket option pricing. J Algorithm Comput Technol 6:101–112

Hull JC (2003) Options, futures and other derivatives. Prentice Hall, Upper Saddle River

Iazzolino G, Fortino A (2012) Credit risk analysis and the KMV Black and Scholes model: a proposal of correction and an empirical analysis. Invest Manag Financ Innov 9:167–181

Jarrow RA, Turnbull SM (2000) Derivative securities, 2nd edn. South-Western College Publishing, Cincinnati

Jasra A, del Moral P (2011) Sequential Monte Carlo methods for option pricing. Stoch Anal Appl 29:292–316

Jeunesse M, Jourdain B (2012) Regularity of the American put option in the Black-Scholes model with general discrete dividends. Stoch Process Appl 122:3101–3125

Kohler M, Krzyzak A, Todorovic N (2010) Pricing of high-dimensional American options by neural networks. Math Financ 20:383–410

Kojima F, Usami T (2012) Nondestructive interrogation of dielectric materials using Markov Chain Monte Carlo method for structural equation model. Stud Appl Electromagn Mech 36:37–44

Lagazio M, Marwala T (2005) Assessing different Bayesian neural network models for militarized interstate dispute. Soc Sci Comput Rev 24(1):1–12

Lajbcygier P (2004) Improving option pricing with the product constrained hybrid neural network. IEEE Trans Neural Netw 15:465–476

Lajbcygier PR, Connor JT (1997) Improved option pricing using artificial neural networks and bootstrap methods. Int J Neural Syst 8:457–471

Leermakers FAM, Bergsma J, Van Der Gucht J (2012) Hybrid Monte Carlo self-consistent field approach to model a thin layer of a polyelectrolyte gel near an adsorbing surface. J Phys Chem A 116:6574–6581

Li M (2011) Research on the mixed method of option price forecasting based on support vector machine. In: Proceedings of the 2nd international conference on artificial intelligence, management science and electronic commerce, Zhengzhou, 2011, pp 6755–6758

Liang X, Zhang H, Yang J (2006) Pricing options in Hong Kong market based on neural networks. Lect Note Comput Sci 4234:410–419

Liang X, Zhang H, Xiao J, Chen Y (2009) Improving option price forecasts with neural networks and support vector regressions. Neurocomputing 72:3055–3065

Liesenfeld R, Richard J (2008) Improving MCMC, using efficient importance sampling. Comput Stat Data Anal 53:272–288

MacKay DJC (1991) Bayesian methods for adaptive models. Ph.D. thesis, California Institute of Technology, Pasadena

MacKay DJC (1992) A practical Bayesian framework for backpropagation networks. Neural Comput 4:448–472

Marwala T (2001) Fault identification using neural networks and vibration data. Ph.D. thesis, University of Cambridge, Cambridge

Marwala T (2007) Bayesian training of neural network using genetic programming. Pattern Recognit Lett 28:1452–1458. doi:org/10.1016/j.patrec.2007.034

Marwala T (2009) Computational intelligence for missing data imputation, estimation and management: knowledge optimization techniques. IGI Global Publications, New York

Marwala T (2010) Finite element model updating using computational intelligence techniques. Springer, London

Marwala T (2012) Condition monitoring using computational intelligence methods. Springer, London

Marwala T, Lagazio M (2011) Militarized conflict modeling using computational intelligence techniques. Springer, London

Merton RC (1973) Theory of rational option pricing. Bell J Econ Manag Sci 4:141–183

Metropolis N, Rosenbluth AW, Rosenbluth MN, Teller AH, Teller E (1953) Equation of state calculations by fast computing machines. J Chem Phys 21:1087

Misev A, Sahpaski D, Pejov L (2012) Implementation of hybrid Monte Carlo (molecular dynamics) quantum mechanical methodology for modeling of condensed phases on high performance computing environment. Adv Intell Soft Comput 150:269–277

Neal RM (1993) Probabilistic inference using Markov chain Monte Carlo methods. University of Toronto technical report CRG-TR-93-1, University of Toronto, Toronto

Niu D-X, Shi H-F, Wu DD (2012) Short-term load forecasting using Bayesian neural networks learned by hybrid Monte Carlo algorithm. Appl Soft Comput J 12:1822–1827

Nqi FZ, Lidouh A (2010) The constrained option pricing with neural networks and isotonic regression. Int J Math Anal 4:201–209

Panayiotis AC, Spiros MH, Chris C (2004) Option pricing and trading with artificial neural networks and advanced parametric models with implied parameters. Proc IEEE Int Conf Neural Netw 4:2741–2746

Pires MM (2005) American option pricing using computational intelligence methods. Masters thesis, University of the Witwatersrand, Johannesburg

Pires MM, Marwala T (2004) American option pricing using multi-layer perceptron and support vector machine. Proc IEEE Int Conf Syst Man Cybern 2:1279–1285

Pires MM, Marwala T (2005) American option pricing using Bayesian multi-layer perceptrons and Bayesian support vector machines. In: Proceedings of the IEEE 3rd international conference on computational cybernetics, Mauritius, 2005, pp 219–224

Robert CP, Casella G (2004) Monte Carlo statistical methods. Springer, London

Ross S, Westerfield RW, Jordan BD, Firer C (2001) Fundamentals of corporate finance, 2nd South African edn. McGraw-Hill Book Company, Sydney

Schölkopf B, Smola AJ (2002) Learning with kernels. MIT Press, Cambridge

Schölkopf B, Burges CJC, Smola AJ (1999) Advances in kernel methods: support vector learning. MIT Press, Cambridge

Seghiouer H, Lidouh A, Nqi FZ (2011) Parallel Monte Carlo method for pricing Asian options using trapezium scheme. Appl Math Sci 5:1533–1547

Shawe-Taylor J, Cristianini N (2004) Kernel methods for pattern analysis. Cambridge University Press, Cambridge

Shen C, Wang X, Yu D (2012) Feature weighting of support vector machines based on derivative saliency analysis and its application to financial data mining. Int J Adv Comput Technol 4:199–206

Steinwart I, Christmann A (2008) Support vector machines. Springer, New York

Thapa N, Ziegler J, Moen C (2012) Existence of optimal parameters for the Black-Scholes option pricing model. Int J Pure Appl Math 78:523–534

Thomaidis NS, Tzastoudis VS, Dounias GD (2007) A comparison of neural network model selection strategies for the pricing of S&P 500 stock index options. Int J Artif Intell Tool 16:1093–1113

Venegas-Martínez F (2005) Bayesian inference, prior information on volatility, and option pricing: a maximum entropy approach. Int J Theor Appl Financ 8:1–12

Verrall RJ, Wüthrich MV (2012) Reversible jump Markov chain Monte Carlo method for parameter reduction in claims reserving. N Am Actuar J 16:240–259

Wang H-W, Huang S-C (2006) Hybrid wavelet-SVMs for modeling derivatives valuation. In: Proceedings of the 9th joint conference on information sciences, Kaohsiung, 2006, art. no. CIEF-134

Wang L, Zhang T, Jia Z, Ding L (2012) Face detection algorithm based on hybrid Monte Carlo method and Bayesian support vector machine. Concurr Comput Pract Exp (Article in press)

Yousuf-Khaliq AQM, Kleefeld B (2012) The numerical approximation of nonlinear Black-Scholes model for exotic path-dependent American options with transaction cost. Int J Comput Math 89:1239–1254

Zhang H, Lin H (2007) Option pricing models based on wavelet neural network. J Southeast Univ (Nat Sci Ed) 37:716–720

Zhang J, Tang WH, Zhang LM, Huang HW (2012) Characterising geotechnical model uncertainty by hybrid Markov chain Monte Carlo simulation. Comput Geotech 43:26–36

Chapter 6
Rough Sets Approach to Economic Modeling: Unlocking Knowledge in Financial Data

Abstract Building models to accurately forecast financial markets has drawn the attention of economists, bankers, mathematicians and scientists alike. The financial markets are the foundation of every economy and there are many aspects that affect the direction, volume, price, and flow of traded stocks. The markets' weakness to external and non-financial features as well as the ensuing volatility makes the development of a robust and accurate financial market forecasting model an interesting problem. In this chapter a rough set theory based forecasting model is applied to the financial markets to identify a set of reducts and possibly a set of trading rules based on trading data.

6.1 Introduction

Rough set theory was proposed by Pawlak (1991) and is a mathematical method which models imprecision and uncertainty. It allows one to approximate sets that are challenging to describe even with available information. In this chapter, rough sets are used to create an intelligent and transparent trading system. The benefits of rough sets, as with many other artificial intelligence approaches, are that they do not require strict *a priori* assumptions about the mathematical relationships of such complex systems, as is usually essential for the multivariate statistical methods (Machowski and Marwala 2005; Marwala 2012; Crossingham and Marwala 2008; Marwala and Lagazio 2011). Rough set theory is based on the assumption that the information of interest is related with some information from its universe of discourse (Tettey et al. 2007; Crossingham et al. 2009; Marwala and Crossingham 2008; Crossingham and Marwala 2008).

Virginia and Nguyen (2013) applied rough sets theory to automatically produce a thesaurus from a corpus. They investigated the relationship between keywords and the sets produced based on rough sets theory. The results they obtained showed that applying rough sets theory is a viable approach to automatically create a thesaurus.

Pérez-Díaz et al. (2012) applied rough sets to filter spam. The proposed method was found to perform better than other anti-spam filtering methods such as support vector machines, Adaboost and Bayes classifiers.

Abed-Elmdoust and Kerachian (2012) applied rough sets for wave height prediction. They applied rough sets to Lake Superior in North America to identify decision rules for wave height prediction. Rough sets approach was found to perform better than support vector machines, Bayesian networks, neural networks, and adaptive neuro-fuzzy inference system on wave height prediction and offered simple decision rules which can used by users.

Villuendas-Rey et al. (2012) applied rough sets for nearest prototype classification. They combined rough set theory with compact sets to obtain a reduced prototype set and obtained good classification accuracy.

Zhou (2012) applied rough sets to build coal mine safety decision system and obtained good results whereas Wang et al. (2006) successfully applied rough set theory to handle uncertainty and, thereby, decreasing the redundancy of evaluating the degree of malignancy in brain glioma, based on magnetic resonance imaging findings as well as the clinical data before an operation.

Further applications of rough set theory consist of the work by Xie et al. (2011) who applied rough sets for land cover retrieval from remote sensing images, Azadeh et al. (2011) who applied a rough set method to evaluate the efficiency of personnel, Salamó and López-Sánchez (2011) who applied rough sets for choosing features in Case-Based Reasoning classifiers, Lin et al. (2011) who applied rough set theory to forecast customer churn in credit card accounts, Zhang et al. (2008) who applied rough sets to control reagents in an ionic reverse flotation process, Huang et al. (2011) who used rough sets in patent development resource allocation, Zou et al. (2011) who applied rough sets for distributor choice in a supply chain management system, Gong et al. (2010) in a rare-earth extraction process, Chen et al. (2010) to diagnose pneumonia in the elderly, Yan et al. (2010) for forecasting soil moisture and Liao et al. (2010) to model brand trust.

The principal notion of rough set theory is an indiscernibility relation, where indiscernibility designates indistinguishable from one another. For knowledge extraction from data with numerical attributes, special approaches are used. The cost commonly used step called *discretization* is applied before the principal step of rule induction or decision tree generation is used (Crossingham and Marwala 2007; Mpanza 2011). A number of methods that have been used for discretization are Boolean reasoning, Equal-Width-Bin (EWB) partitioning and Equal-Frequency-Bin (EFB) partitioning (Jaafar et al. 2006; Fayyad and Irani 1993; Marwala and Lagazio 2011; Marwala 2009; Mpanza and Marwala 2011).

In this chapter we apply rough sets to accurately forecast the financial market. The next chapter describes the theory of rough sets.

6.2 Rough Sets

The main aim of utilizing rough sets is to generate approximations of a number of ideas from the gathered data. Rough set theory has advantages because it does not require (Crossingham 2007; Nelwamondo 2008; Marwala and Lagazio 2011; Marwala 2012):

• any new information about the experimental training data for example the statistical probability; and
• formulation in possibility terms as is done in fuzzy set theory (Pawlak and Munakata 1996).

Rough set theory is intended to approximate sets that are difficult to describe with the information available (Ohrn 1999; Ohrn and Rowland 2000; Marwala and Lagazio 2011; Marwala 2012). It is aimed at classifying imprecise, uncertain, or incomplete information. Two approximations, the upper and lower estimation are formulated to deal with inconsistent information. The data are characterized using an information table.

Rough set theory is based on a set of rules, which are explained in terms of linguistic variables. Rough sets are important aspect of artificial intelligence and have been applied to machine learning and decision analysis, mainly in the analysis of decisions in which there are contradictions. Since they are rule-based, rough sets are highly transparent nonetheless they are not as accurate as other artificial intelligence methods. Nevertheless, they are not good as universal approximators, because other machine learning methods such as neural networks are better in their predictions. Therefore, in machine learning, there is always a trade-off between forecasting accuracy and transparency.

Crossingham and Marwala (2007) offered a method to optimize the partition sizes of rough set using a number of optimization approaches. Three optimization techniques were used to granularize the variables: the genetic algorithm, hill climbing and simulated annealing. These optimization methods maximize the classification accuracy of the rough sets. The three techniques were compared for their computational time, accuracy, and number of rules produced and then applied to an HIV data set. The optimized method results were then compared to a non-optimized discretization technique, using Equal-Width-Bin (EWB) partitioning. The accuracies attained after optimizing the partitions using a genetic algorithm (GA), hill climbing, and simulated annealing (SA) were 66.89, 65.84, and 65.48 %, respectively, compared to the accuracy of the EWB partitioning of 59.86 %. Rough sets gave the plausibility of the estimated HIV status and the linguistic rules relating to how demographic parameters influence the risk of HIV.

Rough set theory enables reasoning from vague and imprecise data (Goh and Law 2003). It is based on the assumption that some observed information is, somehow, related to some information in the universe of the discourse (Komorowski et al. 1999; Yang and John 2006; Kondo 2006; Marwala and Lagazio 2011; Marwala 2012). Objects with the same information are indiscernible in the perspective of the existing information. An elementary set that consists of indiscernible objects forms a basic granule of knowledge. A union of an elementary set is known as a *crisp set*, otherwise, the set is said to be *rough*. In the next sub-sections, rough set theory is explained.

6.2.1   Information System

An information system (Λ), is deemed as a pair (U, A) where U is a finite set of objects known as the universe and A is a non-empty finite set of attributes as illustrated as follows (Crossingham 2007; Yang and John 2006; Nelwamondo 2008; Marwala 2009, 2012; Marwala and Lagazio 2011).

$$\Lambda = (U, A) \tag{6.1}$$

All attributes $a \in A$ have values, which are elements of a set V_a of the attributes a (Dubois 1990; Crossingham 2007; Marwala and Lagazio 2011; Marwala 2009):

$$a : U \to V_a \tag{6.2}$$

A rough set is explained with a set of attributes and the indiscernibility relation between them. Indiscernibility is described in the next subsection.

6.2.2   The Indiscernibility Relation

The *indiscernibility relation* is one of important notions of rough set theory (Grzymala-Busse and Hu 2001; Grzymala-Busse 2004; Grzymala-Busse and Siddhaye2004; Zhao et al. 2007; Pawlak and Skowron 2007; Marwala and Lagazio 2011; Marwala 2012). *Indiscernibility* essentially implies similarity (Goh and Law 2003) and, accordingly, these sets of objects are indistinguishable. Given an information system Λ and subset $B \subseteq A$, B the indiscernibility describes a binary relation $I(B)$ on U such that (Pawlak et al. 1988; Ohrn 1999; Wu et al. 2003; Ohrn and Rowland 2000; Nelwamondo 2008; Marwala and Lagazio 2011; Marwala 2012):

$$(x, y) \in I(B)$$

if and only if

$$a(x) = a(y) \tag{6.3}$$

for all $a \in A$ where $a(x)$ indicates the value of attribute a for element x. Equation 6.3 put forward that any two elements that are elements of $I(B)$ should be indistinguishable from the point of view of a. Let's say that U has a finite set of N objects $\{x_1, x_2, \ldots, x_N\}$. Let Q be a finite set of n attributes $\{q_1, q_2, \ldots, q_n\}$ in the same information system Λ, then (Inuiguchi and Miyajima; 2007; Crossingham 2007; Nelwamondo 2008; Marwala and Lagazio 2011; Marwala 2012):

$$\Lambda = \langle U, Q, V, f \rangle \tag{6.4}$$

where f stands for the total decision function, called the *information function*. From the description of the indiscernibility relation, two objects have a similarity relation to attribute a if they universally have the same attribute values.

6.2.3 Information Table and Data Representation

An information table is used in rough sets theory as a method for indicating the data. Data in the information table are arranged, focused on their condition attributes and decision attributes (D). *Condition attributes* and *decision attributes* are analogous to the independent variables and dependent variable (Goh and Law 2003). These attributes are separated into $C \cup D = Q$ and $C \cup D = 0$. Data is specified in the table and each object is described in an *Information System* (Komorowski et al. 1999).

6.2.4 Decision Rules Induction

Rough sets also necessitate creating decision rules for a given information table. The rules are normally based on condition attributes values (Bi et al. 2003; Slezak and Ziarko 2005). The rules are offered in an '*if CONDITION(S)-then DECISION*' arrangement. Stefanowski (1998) used a rough set method for inference in decision rules.

6.2.5 The Lower and Upper Approximation of Sets

The lower and upper estimates of sets are defined based on the indiscernibility relation. The *lower approximation* is the aggregation of cases whose equivalent classes are restricted in the cases that require to be approximated, while the *upper approximation* is defined as the aggregation of classes that are incompletely contained in the set to be approximated (Rowland et al. 1998; Degang et al. 2006; Witlox and Tindemans 2004). If X is defined as a set of all cases defined by a particular value of the decision and that any finite union of elementary set, related to B called a B-definable set (Grzymala-Busse and Siddhaye 2004) then set X can be approximated by two B-definable sets, called the B-lower approximation indicated by $\underline{B}X$ and B-upper approximation $\overline{B}X$. The B-lower approximation is written as follows (Bazan et al. 2004; Crossingham 2007; Nelwamondo 2008; Marwala and Lagazio 2011; Marwala 2012):

$$\underline{B}X = \{x \in U \,|[x]_B \subseteq X\} \tag{6.5}$$

and the B-upper approximation is written as follows (Crossingham 2007; Nelwamondo 2008; Marwala and Lagazio 2011; Marwala 2012):

$$\overline{B}X = \{x \in U \,|[x]_B \cap X \neq 0\} \tag{6.6}$$

There are other methods that have been explained for defining the lower and upper approximations for an entirely detailed decision table and these include estimating the lower and upper approximation of X using Eqs. 6.7 and 6.8, as follows (Grzymala-Busse 2004; Crossingham 2007; Nelwamondo 2008; Marwala and Lagazio 2011; Marwala 2012):

$$\cup \{[x]_B \,|x \in U, [x]_B \subseteq X\} \tag{6.7}$$

$$\cup \{[x]_B \,|x \in U, [x]_B \cap X \neq 0\} \tag{6.8}$$

The definition of definability is revised in situations of incompletely specified tables. In this case, any finite union of characteristic sets of B is called a *B-definable set*. Three different definitions of approximations have been discussed by Grzymala-Busse and Siddhaye (2004). By letting B be a subset of A of all attributes and $R(B)$ be the characteristic relation of the incomplete decision table with characteristic sets $K(x)$, where $x \in U$, the following can be defined (Grzymala-Busse 2004; Crossingham 2007; Nelwamondo 2008; Marwala and Lagazio 2011):

$$\underline{B}X = \{x \in U \,|K_B(x) \subseteq X\} \tag{6.9}$$

and

$$\overline{B}X = \{x \in U \,|K_B(x) \cap X \neq 0\} \tag{6.10}$$

Equations 6.9 and 6.10 are called as *singletons*. The subset lower and upper approximations of incompletely quantified data sets can then be written as follows (Nelwamondo 2008; Marwala and Lagazio 2011; Marwala 2012):

$$\cup \{K_B(x) \,|x \in U, K_B(x) \subseteq X\} \tag{6.11}$$

and

$$\cup \{K_B(x) \,|\, x \in U, K_B(x) \cap X = 0\} \qquad (6.12)$$

More information on these methods can be accessed in (Grzymala-Busse and Hu 2001; Grzymala-Busse and Siddhaye 2004; Crossingham 2007; Marwala and Lagazio 2011; Marwala 2012). It can be inferred from these properties that a crisp set is only defined if $\underline{B}(X) = \overline{B}(X)$. *Roughness* is, therefore, defined as the difference between the upper and the lower approximation.

6.2.6 Set Approximation

Several properties of rough sets have been identified by Pawlak (1991). A significant property of rough set theory is the definability of a rough set (Quafafou 2000). This was described for the condition when the lower and upper approximations are equal. If this is not the condition, then the set is un-definable. As described by Marwala (2012) various distinctive cases of definability are (Pawlak et al. 1988; Crossingham 2007; Nelwamondo 2008; Marwala 2009, 2012; Marwala and Lagazio 2011):

- *Internally definable* set: Here, $\underline{B}X \neq 0$ and $\overline{B}X = U$. The attribute set B has objects that definitely are elements of the target set X, even though there are no objects that can definitively be excluded from the set X.
- *Externally definable* set: Here, $\underline{B}X = 0$ and $\overline{B}X \neq U$. The attribute set B has no objects that certainly are elements of the target set X, even though there are objects that can definitively be excluded from the set X.
- *Totally un-definable* set: Here, $\underline{B}X = 0$ and $\overline{B}X = U$. The attribute set B has no objects that, definitely, are elements of the target set X, even though there are no objects that can, definitively, be excluded from the set X.

6.2.7 The Reduct

An added property of rough sets is the *reduct* which is a notion that expresses whether there are attributes B in the information system that are more important to the knowledge characterized in the equivalence class structure than other attributes. It is important to identify whether there is a subset of attributes which could be totally explained by the knowledge in the database. This attribute set is called the *reduct*.

Terlecki and Walczak (2007) defined the relations between rough set reducts and emerging patterns and established a practical application for these observations for the minimal reduct problem, using these to test the differentiating factor of an attribute set. Shan and Ziarko (1995) properly defined a *reduct* as a subset of attributes $RED \subseteq B$ such that:

- $[x]_{RED} = [x]_B$. To be precise, the equivalence classes that were induced by reducing the attribute set RED are equal to the similar class structure that was induced by the full attribute set B.
- Attribute set RED is minimal for the reason that $[x]_{(RED-A)} \neq [x]_B$ for any attribute $A \in RED$. Basically, there is no attribute that can removed from the set RED without changing the equivalent classes $[x]_B$.

Consequently a reduct can be pictured as an appropriate set of characteristics that can, sufficiently, define the category's structure. One property of a reduct in an information system is that it is not unique since there may be other subsets of attributes which may also conserve the equivalence class structure in the information system. The set of features that are present in all reducts is a *core*.

6.2.8   Boundary Region

As described by Marwala (2012) and many researchers before that, the *boundary region*, which is defined as the difference $\overline{B}X - \underline{B}X$, is a region which is composed of objects that cannot be included nor excluded as elements of the target set X. Basically, the lower approximation of a target set is an approximation which consists only of those objects which can be identified as elements of the set. The upper approximation is a rough estimate and includes objects that may be elements of the target set. The boundary region is the area between the upper and lower approximation.

6.2.9   Rough Membership Functions

A *rough membership function* is a function $\mu_A^x : U \rightarrow [0, 1]$ that, when used to object x, estimates the degree of intersection between set X and the indiscernibility set to which x is an element of. The rough membership function is used to approximate the plausibility and can be written as follows (Pawlak 1991; Crossingham 2007; Nelwamondo 2008; Marwala and Lagazio 2011; Marwala 2012):

$$\mu_A^x(X) = \frac{|[x]_B \cap X|}{|[x]_B|} \qquad (6.13)$$

The rough membership function is analogous to the fuzzification process. The significant characteristic of a rough membership function is that it is derived from data (Hoa and Son 2008; Crossingham 2007).

6.3 Discretization Methods

The approaches which permit continuous data to be processed encompass discretization. There are a number of approaches that can be discretized and these include Equal-Frequency-Bin (EFB) partitioning, Boolean reasoning, entropy and Naïve method (Crossingham 2007).

6.3.1 Equal-Frequency-Bin (EFB) Partitioning

EFB partitioning positions the values of every attribute in increasing arrangements and splits them into k bins where given m cases every bin has m/k neighboring values. In most cases, repeated values will possibly occur. The EFB partitioning can be applied as follows (Crossingham and Marwala 2007; Grzymala-Busse 2004; Crossingham 2007; Marwala and Lagazio 2011; Marwala 2012):

- Position the values of every attribute $\left(v_1^a, v_2^a, v_3^a, \ldots, v_m^a\right)$ into intervals whereby m is the number of cases.
- Consequently every interval is made of the following successive values:

$$\lambda = \frac{m}{4} \qquad (6.14)$$

- The cut-off points may be calculated by means of the this equation which is true for $i = 1,2,3$ where k intervals can be estimated for $i = 1, \ldots, k-1$:

$$c_i = \frac{v_{i\lambda} + v_{i\lambda+1}}{2} \qquad (6.15)$$

6.3.2 Boolean Reasoning

As proposed by Nguyen and Skowron (1995) and also explained by Marzuki and Ahmad (2007), the Boolean reasoning discretization operates by identifying the minimum partition P_a of V_a for all $a \in A$ such that:

$$\Lambda^P = \left(U, A^P \cup D\right) \qquad (6.16)$$

is still a consistent information system given a consistent decision system:

$$\Lambda = (U, A \cup D) \qquad (6.17)$$

More information on Boolean reasoning and its applications can be found in Pawlak and Skowron (2007) as well as Nguyen (2005, 2006).

6.3.3 Entropy Based Discretization

As described by Doughety et al. 1995 entropy based discretization method iteratively divide the value of each attribute with the purpose of ensuring that the local quantity of entropy is optimized. The stopping criterion is the minimum description length principle (Doughety et al. 1995). The gain cut b is written as follows (Doughety et al. 1995):

$$U_0 = \{x \in U \,|a(x) < b\} \tag{6.18}$$

$$U_1 = \{x \in U \,|b < a(x)\} \tag{6.19}$$

$$E(U) = \sum_{i=1}^{l} P\,(d_i\,|U\,)\log_2 P\,(d_i\,|U\,) \tag{6.20}$$

$$Gain(a,b,U) = E(U) - \left(\frac{|U_0|}{U}E\,(U_0) + \frac{|U_1|}{U}E\,(U_1)\right) \tag{6.21}$$

b is accepted if (Doughety et al. 1995):

$$Gain(a,b,U) > \frac{\log_2\,(|U|-1)}{|U|} + \frac{\log\,(3^l - 2) - (l E(U) - I_0 E\,(U_0) - I_1 E\,(U_1)}{|U|} \tag{6.22}$$

where l is number of decision class in U. The entropy based discretization has been successfully applied for hierarchical clustering (Cao et al. 2012) and forecasting the stock market (Chen et al. 2010). More information on entropy based discretization are described by Huang et al. (2009) as well as Sheri and Corne (2009).

6.3.4 Naïve Algorithm

This algorithm makes a cut between two neighbouring points if they have different classes. If the sorted values of b is $v_1^b < \cdots < v_i^b < \cdots < V_m^b$ where $\partial_j^b = \left\{dx\,|b(x) = v_j^b\right\}$ then (Ohrn 1999);

$$c_b = \left\{\frac{v_j^b + v_{j+1}^b}{2}\,\middle|\,\partial_j^b \neq \partial_{j+1}^b \ and \ j = 1,\ldots,m-1\right\} \tag{6.23}$$

6.4   Rough Set Formulation

Rough set modeling is categorized into these five stages (Grzymala-Busse 2004; Crossingham 2007; Marwala and Lagazio 2011; Marwala 2012):

1. Choose the data.
2. Pre-process the data by discretizing the data and removing outliers.
3. If reducts are factored into account, use the cleaned data to produce reducts. A *reduct* is the most succinct mode in which objects can be discerned. Therefore, a reduct is the minimal subset of attributes that classify elements of the universe to the same extent as the whole set of attributes. Define lower and upper approximations to handle contradictions.
4. Rules are extracted or generated based on condition attribute values.
5. Test the newly generated rules on a test set.

The technique for formulating rough sets and extracting rules is given in Algorithm 6.1 (Crossingham 2007; Marwala 2012; Khoza and Marwala 2012). After the rules have been extracted, they can be tested using a set of testing data. The standard rough set implementation is indicated in Fig. 6.1 (Khoza and Marwala 2011).

Algorithm 6.1 Procedure to generate a rough set model (Crossingham 2007)

Input:	Condition and Decision Attributes
Output:	Certain and Possible Rules

1   Obtain the data set to be used;
2   **Repeat**
3   **for** *conditional_attribute* ← 1 to *size_of_training_data* **do**
4   Pre-process data to ensure that it is ready for analysis;
5   Discretize the data according to the optimization technique;
6   Compute the lower approximation, as defined in equation 6.5;
7   Compute the upper approximation, as defined in equation 6.6;
8   From the general rules, calculate plausibility measures for an object x belonging to set X, as defined by equation 6.13;
9   Extract the *certain* rules from the lower approximation generated for each subset;
10   Similarly, extract the *possible* rules from the upper approximation of each subset;
11   Remove the generated rules for the purposes of testing on unseen data;
12   Compute the classifier performance using the AUC;
13   **End**
14 **until** Optimization technique termination condition;

Fig. 6.1 The rough set
predictive model

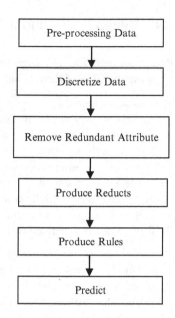

6.5   Application to Modeling the Stock Market

In this chapter we model the Johannesburg Stock Exchange's All Share Index
(ALSI) from 2006 to 2011 as was done by Khoza and Marwala (2011). The data
was randomly divided into 75 and 25 % ratio. Seventy-five percent of the data was
used for training and 25 % was used for validation. There was a total of ten attributes
used and these are shown Table 6.1 (Khoza 2012).

The decision attribute in this chapter can be represented as follows (Khoza 2012;
Al-Qaheri et al. 2008):

$$D = \frac{\sum_{i=1}^{i=n} ((n+1) - i) \times \|Close(i) - Close(i-1)\|}{\sum_{n=1}^{n} i} \tag{6.24}$$

The decision attribute D is Eq. 6.24 is normalized to limit its value to fall between
-1 and $+1$. A value of $+1$ demonstrates that every day for the next n days into the
future the price closes higher than today and likewise a value of -1 shows that every
day for the next n days into the future the price closes lower than it is at present.
In essence D is an indicator on whether we buy or sell and it indicates the future
direction of the index.

When rough set technique was implemented to model the stock market, as
described above, the dominance of each attribute as measured as an appearance
percentage of the total number of reducts generated is shown in Table 6.2 (Khoza
and Marwala 2011).

Table 6.1 Table of attributes

Attributes	Description
Open	The opening price
High	The highest registered price on the day
Low	The lowest registered price on the day
Close	The closing price
Adjusted close	The adjusted closing price
Moving average	Moving average over 5 days
Momentum	$P_i - P_{i-4}$
Rate of change (ROC)	$\dfrac{momentum}{P_i} \times 100$
Disparity	$\dfrac{P_i}{Moving\ Average} \times 100$
Decision (D)	Decision attribute

Table 6.2 Attribute used to model the stock market

Attribute	Count	Percentage (%)
Open	54	29.35
High	73	39.67
Low	81	44
Close	71	38.58
Adj Close	54	29.34
MAV	68	36.95
Momentum	72	39.14
ROC	69	37.5
Disparity	89	48.37

In essence, the rough set procedure applied to model the stock market is oulined in Fig. 6.2. As identified by Khoza and Marwala (2011), in this chapter we also identify that the six most important attributes identified from the rough sets are the highest registered price on the day, followed by the lowest registered price on the day, which is in turn followed by the closing price of the day, followed by the momentum, which is followed by the rate of change of price which is followed by the disparity. To provide comparison, rules were built using the 182 reducts created in the first stage of the previous phase and then a second set of rules was extracted using only the core reduct. With 182 reducts a total of 1,004 rules were created and with the core reduct above a total of 246 rules were generated.

The decision table is built by having columns as several technical indicators and the rows indicating trading data at each point in time, whereas the window offers a "snapshot" of the state of the market in that period. There are several parameters that influence the accuracy of the model these are the data split ratio, discretization algorithm and the classifier technique (Doughety et al. 1995). To build the most robust model the correct combination of these parameters is necessary.

Because it was observed in (Doughety et al. 1995) that the quality of the model its accuracy depends heavily on the discretization algorithm, a number of discretization algorithms were used so as to assess which gave the best results. These were: EFB,

Fig. 6.2 Prediction method

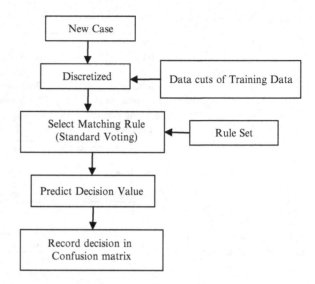

Table 6.3 Discretization algorithm accuracy comparison

	Evaluation criteria		
Discretization algorithm	Reducts	Rules	Accuracy (%)
EFB (with four data cuts)	182	1,004	86.8
BR	2	1,510	57.7
Entropy	2	484	64.5
Naïve algorithm	32	31,188	43

Table 6.4 EFB cuts comparison

	Evaluation criteria		
EFB cuts	Reducts	Rules	Accuracy (%)
3	171	943	66.0
4	182	1,004	86.8
5	190	1,856	77.3
6	197	2,450	76.1

the Boolean Reasoning (BR) algorithm, Entropy/MDL algorithm, and the Naïve algorithm. The results obtained when these algorithms were implemented are shown in Table 6.3.

From the results in Table 6.3, it is observed that the EFB gave the best results and, therefore, all analysis in this chapter will be based on the EFB. The relationship between the number of data cuts and accuracy when the EFB implemented is shown in Table 6.4.

As shown in Table 6.4, it was observed that 4 data cuts offered the best results, from which other tuning of parameters can be based. Using the standard voting classifier, the difference in accuracy and rules generated from the EFB (four data cuts) with the normal reducts and the core reducts are shown in Table 6.5.

Table 6.5 Normal and core reduct comparison

Reducts	Rules	Accuracy (%)
Normal	1,004	86.8
Core reduct	246	80.4

Table 6.6 Rough set model confusion matrix

		Predicted		
		0	1	
Actual	0	147	14	0.9130
	1	43	87	0.6692
		0.7736	0.8613	**0.8041**

Table 6.5 demonstrates a trade-off between the number of rules generated and the accuracy of the model. Whilst we aim to make the forecasting model as small as possible, the accuracy of the model needs to be factored into account. Rough set model's confusion matrix is shown in Table 6.6 for the core reduct based system.

6.6 Conclusion

A rough set theory based predictive model for stock prices was proposed in this chapter. The data was randomly divided into a 75 %/25 % which were the training and validation data sets, respectively. The data set was discretized using Equal frequency bin, entropy approach, Boolean reasoning and Naive method. The models showed a high degree of accuracy on classifying the daily movements of the Johannesburg Stock Exchange's All Share Index and that the EFB gave the best results.

References

Abed-Elmdoust A, Kerachian R (2012) Wave height prediction using the rough set theory. Ocean Eng 54:244–250

Al-Qaheri H, Zamoon S, Hassanien A (2008) Rough set generating prediction rules for stock price movement. Int Res J Finance Econ 29:70–80

Azadeh A, Saberi M, Moghaddam RT, Javanmardi L (2011) An integrated data envelopment analysis – artificial neural network-rough set algorithm for assessment of personnel efficiency. Expert Syst Appl 38:1364–1373

Bazan J, Nguyen HS, Szczuka M (2004) A view on rough set concept approximations. Fundamenta Informaticae 59:107–118

Bi Y, Anderson T, McClean S (2003) A rough set model with ontologies for discovering maximal association rules in document collections. Knowl Based Syst 16:243–251

Cao B-W, Xue Q, Li A-L (2012) Attribute discretization algorithm for data stream based on hierarchical clustering and information entropy. In: Proceedings of the 2nd international conference on consumer electronics, communications and networks, Three Gorges, China, pp 193–196

Chen Y, Miao D, Wang R (2010) A rough set approach to feature selection based on ant colony optimization. Pattern Recognit Lett 31(3):226–233

Crossingham B (2007) Rough set partitioning using computational intelligence approach. M.Sc. thesis, University of the Witwatersrand, Johannesburg

Crossingham B, Marwala T (2007) Using optimisation techniques to granulise rough set partitions. Comput Model Life Sci 952:248–257

Crossingham B, Marwala T (2008) Using optimisation techniques for discretizing rough set partitions. Int J Hybrid Intell Syst 5:219–236

Crossingham B, Marwala T, Lagazio M (2009) Evolutionarily optimized rough set partitions. ICIC Express Lett 3:241–246

Degang C, Wenxiu Z, Yeung D, Tsang ECC (2006) Rough approximations on a complete completely distributive lattice with applications to generalized rough sets. Inf Sci 176:1829–1848

Doughety J, Kohavi R, Sahami M (1995) Supervised and unsupervised discretization of continous features. In: Proceedings of the XII international conference on machine learning, Lake Tahoe, 1995, pp 2940–2301

Dubois D (1990) Rough fuzzy sets and fuzzy rough sets. Int J Gen Syst 17:191–209

Fayyad U, Irani K (1993) Multi-interval discretization of continuous valued attributes for classification learning. In: Proceedings of the 13th international joint conference on artificial intelligence, Chambéry, 1993, pp 1022–1027

Goh C, Law R (2003) Incorporating the rough sets theory into travel demand analysis. Tour Manag 24:511–517

Gong J, Yang H, Zhong L (2010) Case-based reasoning based on rough set in rare-earth extraction process. In: Proceedings of the 29th Chinese control conference, Beijing, 2010, pp 70–1706

Grzymala-Busse JW (2004) Three approaches to missing attribute values – a rough set perspective. In: Proceedings of the IEEE 4th international conference on data mining, Brighton, 2004, pp 57–64

Grzymala-Busse JW, Hu M (2001) A comparison of several approaches to missing attribute values in data mining. Lect Note Artif Intell 205:378–385

Grzymala-Busse JW, Siddhaye S (2004) Rough set approaches to rule induction from incomplete data. In: Proceedings of the 10th international conference on information processing and management of uncertainty in knowledge-based systems, Perugia, 2004, vol 2, pp 923–930

Hoa NS, Son NH (2008) Rough set approach to approximation of concepts from taxonomy. http://logic.mimuw.edu.pl/publikacje/SonHoaKDO04.pdf. Last accessed 9 July 2011

Huang Q, Chen G, Liu Y, Guo W, Fang X (2009) A research and application of continuous attributes discretization based on cloud model and information entropy. In: Proceedings of the 6th international conference on fuzzy systems and knowledge discovery, Tianjin, 2009, pp 34–38

Huang C-C, Liang W-Y, Shian-Hua L, Tseng T-L, Chiang H-Y (2011) A rough set based approach to patent development with the consideration of resource allocation. Expert Syst Appl 38:1980–1992

Inuiguchi M, Miyajima T (2007) Rough set based rule induction from two decision tables. Eur J Oper Res 181:1540–1553

Jaafar AFB, Jais J, Hamid MHBHA, Rahman ZBA, Benaouda D (2006) Using rough set as a tool for knowledge discovery in DSS. In: Proceedings of the 4th international conference on multimedia and information and communication technologies in education, Seville, 2006, pp 1011–1015

Khoza MS (2012) Economic modeling using computational intelligence techniques. Master of Engineering thesis, University of Johannesburg, Johannesburg

Khoza M, Marwala T (2011) A rough set theory based predictive model for stock prices. In: Proceedings of the IEEE 12th international symposium on computational intelligence and informatics, Budapest, 2011, pp 57–62

Komorowski J, Pawlak Z, Polkowski L, Skowron A (1999) A rough set perspective on data and knowledge. In: Klösgen W, Zytkow JM (eds) The handbook of data mining and knowledge discovery. Oxford University Press, New York

Kondo M (2006) On the structure of generalized rough sets. Inf Sci 176:589–600

Liao SH, Chen YJ, Chu PH (2010) Rough-set-based association rules applied to brand trust evaluation model. Lect Note Comput Sci 6443:634–641

Lin CS, Tzeng GH, Chin YC (2011) Combined rough set theory and flow network graph to predict customer churn in credit card accounts. Expert Syst Appl 38:8–15

Machowski LA, Marwala T (2005) Using object oriented calculation process framework and neural networks for classification of image shapes. Int J Innov Comput Inf Control 1:609–623

Marwala T (2009) Computational intelligence for missing data imputation, estimation and management: knowledge optimization techniques. IGI Global Publications, New York

Marwala T (2012) Condition monitoring using computational intelligence methods. Springer, London

Marwala T, Crossingham B (2008) Neuro-rough models for modelling HIV. In: Proceedings of the IEEE international conference on systems, man, and cybernetics, Singapore, pp 3089–3095

Marwala T, Lagazio M (2011) Militarized conflict modeling using computational intelligence techniques. Springer, London

Marzuki Z, Ahmad F (2007) Data mining discretization methods and performances. In: Proceedings of the international conference on electrical engineering and informatics, Badung, 2007, pp 535–537

Mpanza LJ (2011) A rough set approach to bushings fault detection. Master of Engineering thesis, University of Johannesburg, Johannesburg

Mpanza LJ, Marwala T (2011) Artificial neural network and rough set for HV bushings condition monitoring. In: Proceedings of the 15th IEEE international conference on intelligent engineering systems, Kaiserslautern, 2011, arXiv:1108.4618

Nelwamondo FV (2008) Computational intelligence techniques for missing data imputation. Ph.D. thesis, University of the Witwatersrand, Johannesburg

Nguyen HS (2005) Approximate Boolean reasoning approach to rough sets and data mining. Lect Note Comput Sci 3642:12–22

Nguyen HS (2006) Approximate Boolean reasoning: foundations and applications in data mining. Lect Note Comput Sci 4100:334–506

Nguyen HS, Skowron A (1995) Quantization of real-valued attributes. In: Proceedings second international joint conference on information sciences, Wrightsville Beach, 1995, pp 34–37

Ohrn A (1999) Discernibility and rough sets in medicine: tools and applications. Unpublished Ph.D. thesis, Norwegian University of Science and Technology, Trondheim

Ohrn A, Rowland T (2000) Rough sets: a knowledge discovery technique for multifactorial medical outcomes. Am J Phys Med Rehabil 79:100–108

Pawlak Z (1991) Rough sets – theoretical aspects of reasoning about data. Kluwer Academic Publishers, Dordrecht

Pawlak Z, Munakata T (1996) Rough control application of rough set theory to control. In: Proceedings of the 4th European congress on intelligence techniques and soft computing, Aachen, 1996, pp 209–218

Pawlak Z, Skowron A (2007) Rough sets and Boolean reasoning. Inf Sci 177:41–73

Pawlak Z, Wong SKM, Ziarko W (1988) Rough sets: probabilistic versus deterministic approach. Int J Man Mach Stud 29:81–95

Pérez-Díaz N, Ruano-Ordás D, Méndez JR, Gálvez JF, Fdez-Riverola F (2012) Rough sets for spam filtering: selecting appropriate decision rules for boundary e-mail classification. Appl Soft Comput J 12:3671–3682

Quafafou M (2000) α-RST: a generalization of rough set theory. Inf Sci 124:301–316

Rowland T, Ohno-Machado L, Ohrn A (1998) Comparison of multiple prediction models for ambulation following spinal cord injury. In Chute 31:528–532

Salamó M, López-Sánchez M (2011) Rough set based approaches to feature selection for case-based reasoning classifiers. Pattern Recognit Lett 32:280–292

Shan N, Ziarko W (1995) Data-based acquisition and incremental modification of classification rules. Comput Intell 11:357–370

Sheri G, Corne DW (2009) Evolutionary optimization guided by entropy-based discretization. Lect Note Comput Sci 5484:695–704

Slezak D, Ziarko W (2005) The investigation of the Bayesian rough set model. Int J Approx Reason 40:81–91

Stefanowski J (1998) On rough set based approaches to induction of decision rules. In: Polkowski L, Skowron A (eds) Rough sets in knowledge discovery 1: methodology and applications. Physica-Verlag, Heidelberg

Terlecki P, Walczak K (2007) On the relation between rough set reducts and jumping emerging patterns. Inf Sci 177:74–83

Tettey T, Nelwamondo FV, Marwala T (2007) HIV data analysis via Rule extraction using rough sets. In: Proceedings of the 11th IEEE international conference on intelligent engineering systems, Budapest, 2007, pp 105–110

Villuendas-Rey Y, Caballero-Mota Y, García-Lorenzo MM (2012) Using rough sets and maximum similarity graphs for nearest prototype classification. Lect Note Comput Sci 7441:300–307

Virginia G, Nguyen HS (2013) Investigating the potential of rough sets theory in automatic thesaurus construction. Lect Note Electr Eng 157:107–112

Wang W, Yang J, Jensen R, Liu X (2006) Rough set feature selection and rule induction for prediction of malignancy degree in brain glioma. Comput Method Progr Biomed 83:147–156

Witlox F, Tindemans H (2004) The application of rough sets analysis in activity based modelling: opportunities and constraints. Expert Syst Appl 27:585–592

Wu W, Mi J, Zhang W (2003) Generalized fuzzy rough sets. Inf Sci 151:263–282

Xie F, Lin Y, Ren W (2011) Optimizing model for land use/land cover retrieval from remote sensing imagery based on variable precision rough sets. Ecol Model 222:232–240

Yan W, Liu W, Cheng Z, Kan J (2010) The prediction of soil moisture based on rough set-neural network model. In: Proceedings of the 29th Chinese control conference, Beijing, 2010, pp 2413–2415

Yang Y, John R (2006) Roughness bound in set-oriented rough set operations. In: Proceedings of the IEEE international conference on fuzzy systems, Vancouver, 2006, pp 1461–1468

Zhang L, Ji SM, Xie Y, Yuan QL, Zhang YD, Yao ZN (2008) Intelligent tool condition monitoring system based on rough sets and mathematical morphology. Appl Mech Mater 10–12:722–726

Zhao Y, Yao Y, Luo F (2007) Data analysis based on discernibility and indiscernibility. Inf Sci 177:4959–4976

Zhou T (2012) Application of data mining in coal mine safety decision system based on rough Set. Lect Note Comput Sci 7389:34–41

Zou Z, Tseng T-L, Sohn H, Song G, Gutierrez R (2011) A rough set based approach to distributor selection in supply chain management. Expert Syst Appl 38:106–115

Chapter 7
Missing Data Approaches to Economic Modeling: Optimization Approach

Abstract This chapter introduces an auto-associative network with optimization methods for modelling economic data. This resulting architecture is a missing data estimation technique, and this is used to predict the production volume by treating it as a missing variable. The autoassociative network is created using a multi-layered perceptron network, while the optimization techniques which are implemented are particle swarm optimization, genetic algorithms and simulated annealing. The results obtained are then compared.

7.1 Introduction

In this chapter, inference is treated as a correlation phenomenon by applying the auto-associative multi-layer perceptron network (Marwala 2009). This, in essence, yields a missing data estimation problem. Moreau et al. (2012) applied this approach for estimating missing data in the life cycle inventory of hydroelectric power plants, while Tsai and Yang (2012) applied neural networks to improve measurement invariance assessments in survey research data that had some values missing. Kim and Shin (2012) applied the factoring likelihood technique for non-monotone missing data estimation while Rey-del-Castillo and Cardeñosa (2012) applied fuzzy min-max neural networks for missing data imputation.

The missing data framework implemented in this chapter is constructed using a multi-layered perceptron, and the missing data is estimated using three optimization methods; namely; particle swarm optimization, genetic algorithm, and simulated annealing (Marwala 2010, 2012; Marwala and Lagazio 2011). The developed framework is then tested on manufacturing data from the South African Reserve Bank. The next section describes the missing data estimation framework.

T. Marwala, *Economic Modeling Using Artificial Intelligence Methods*, Advanced Information and Knowledge Processing, DOI 10.1007/978-1-4471-5010-7_7,
© Springer-Verlag London 2013

7.2   Missing Data Estimation Method

The missing data estimation procedure suggested in this chapter involves the application of a neural network model that is trained to recall itself (i.e. predict its input vector) and is called an auto-associative neural network (Miranda et al. 2012; Makki and Hosseini 2012). Mathematically, the auto-associative model can be written as follows (Marwala 2009):

$$\{Y\} = f\left(\{X\}, \{W\}\right) \tag{7.1}$$

In Eq. 7.1, $\{Y\}$ is the output vector, $\{X\}$ the input vector and $\{W\}$ is the free parameter vector. In the case of a neural network, the free parameters are called weights. Because the model is trained to predict its own input vector, the input vector $\{X\}$ is approximately equal to output vector $\{Y\}$ and consequently $\{X\} \approx \{Y\}$. In actual fact, the input vector $\{X\}$ and output vector $\{Y\}$ will not always be perfectly the same, therefore an error function expressed as the difference between the input and output vector is defined (Marwala 2009):

$$\{e\} = \{X\} - \{Y\} \tag{7.2}$$

Substituting the value of $\{Y\}$ from Eq. 7.1 into Eq. 7.2, the following expression is obtained (Marwala 2009):

$$\{e\} = \{X\} - f\left(\{X\}, \{W\}\right) \tag{7.3}$$

Because the aim is for the error to be minimized and non-negative, the function can be modified as a square of Eq. 7.3 (Marwala 2009):

$$\{e\} = \left(\{X\} - f\left(\{X\}, \{W\}\right)\right)^2 \tag{7.4}$$

For missing data, some of the values for the input vector $\{X\}$ are not obtainable. Therefore, we can classify the input vector elements into $\{X\}$ known vector represented by $\{X_k\}$ and $\{X\}$ unknown vector represented by $\{X_u\}$. Modifying Eq. 7.4 in terms of $\{X_k\}$ and $\{X_u\}$ we have (Marwala 2009):

$$\{e\} = \left(\left\{\begin{matrix} \{X_k\} \\ \{X_u\} \end{matrix}\right\} - f\left(\left\{\begin{matrix} \{X_k\} \\ \{X_u\} \end{matrix}\right\}, \{W\}\right)\right)^2 \tag{7.5}$$

The error vector in Eq. 7.5 can be condensed into a scalar by integrating over the size of the input vector and the number of training examples as follows (Marwala 2009):

$$E = \left\| \left(\left\{\begin{matrix} \{X_k\} \\ \{X_u\} \end{matrix}\right\} - f\left(\left\{\begin{matrix} \{X_k\} \\ \{X_u\} \end{matrix}\right\}, \{W\}\right)\right) \right\| \tag{7.6}$$

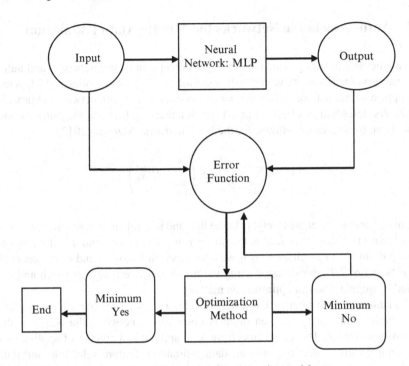

Fig. 7.1 Schematic representation of the missing data estimation model

The objective function expressed in Eq. 7.6 is known as the missing data estimation equation. To estimate the missing input values, Eq. 7.6 is minimized and, in this chapter, artificial intelligence techniques called particle swarm optimization (Kennedy and Eberhart 1995, 2001; Shi and Eberhart 1998; Kennedy 1997) and simulated annealing (Kirkpatrick et al. 1983; Černý 1985; Metropolis et al. 1953; Granville et al. 1994) are applied. It must be taken into account that any optimization technique or a combination of these can be applied to realize this objective. Particle swarm optimization and simulated annealing are selected for the reason that they both have a higher probability of identifying the global optimum solution than traditional optimization techniques such as the scaled conjugate gradient technique, which was used for training the MLP network in Chaps. 3 and 4. For the minimization of Eq. 7.6 to be successful, the identification of a global optimum solution, as opposed to local one, is unequivocally critical because if this is not attained, then a wrong approximation of the missing data will be realized. The missing data process described in this section is illustrated in Fig. 7.1 (Marwala 2009).

Briefly, the objective function known as the missing data estimation equation is derived from the error function of the input and output vector achieved from the trained neural network. The missing data estimation equation is then minimized using the particle swarm optimization method, genetic algorithm and simulated annealing to estimate the missing variables given the observed variables $\{X_k\}$ and the model f explaining the interrelationships and the rules describing the data.

7.3 Auto-associative Networks for Missing Data Estimation

The mathematical background to multilayer perceptron neural networks and auto-associative networks are explained in this section. This chapter applies multi-layered perceptron neural networks to construct auto-associative neural networks (Marwala 2012). As described in Chaps. 3 and 4, the relationship between the output y and input x can be written as follows, for the MLP network (Marwala 2012):

$$y_k = \sum_{j=0}^{M} w_{kj}^{(2)} \tanh \left(\sum_{i=0}^{d} w_{ji}^{(1)} x_i \right) \qquad (7.7)$$

where $w_{ji}^{(1)}$ and $w_{kj}^{(2)}$ denotes weights in the first and second layer, respectively, going from input i to hidden unit j, M is the number of hidden units, and d is the number of output units. In this chapter, as it was described in Chaps. 3 and 4, the network weights in Eq. 7.7 are estimated using the maximum-likelihood approach and the scaled conjugate gradient optimization method.

An auto-associative network is a network that is trained to remember its inputs. This implies that, every time an input is given to the network, the output is the approximated input. These networks have been applied in a number of applications including novelty detection, missing data estimation, feature selection, and data compression (Kramer 1992; Marwala 2009).

There has been more interest in treating the missing data problem by approximation or imputation (Abdella 2005; Abdella and Marwala 2005, 2006; Nelwamondo and Marwala 2007; Nelwamondo 2008). The mixture of the auto-associative neural network and genetic algorithm has been shown to be a successful technique to approximate missing data. The method for estimating missing data in this chapter depends on the identification of the relationships or correlations between the variables that make up the dataset, and the multi-layered perceptron is able to achieve this (Kramer 1992).

Other successful applications of auto-associative network includes its use in fault detection in turbine blades (Lemma and Hashim 2012; Dervilis et al. 2012; Palmé et al. 2011), face recognition (Wang and Yang 2011) and speech recognition (Sivaram et al. 2010).

It must be noted that, on using auto-associative neural networks for data compression, the network has fewer nodes in the hidden layer. Nonetheless, for missing data estimation it is vital that the network is as accurate as possible and that this accuracy is not necessarily achieved through few hidden nodes as is the case when these networks are used for data compression. The auto-associative network is shown in Fig. 7.2 (Marwala 2009).

In this chapter, global optimum techniques, particle swarm optimization and simulated annealing were applied to identify the global optimum solution and are the subject of the next two sections.

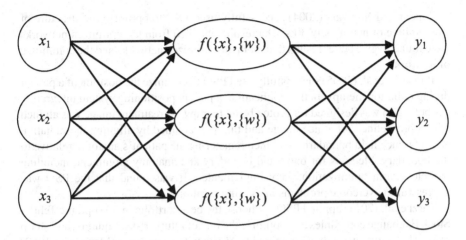

Fig. 7.2 An auto-associative MLP network having two layers of adaptive weights

7.4 Particle Swarm Optimization

This chapter applies particle swarm optimization (PSO) to solve Eq. 7.6. PSO is a stochastic, population-based evolutionary procedure that has been extensively used for the optimization of complex problems (Engebrecht 2005). It is inspired by principles that are based on swarm intelligence. Swarm intelligence consists of two aspects and these are: group knowledge and individual knowledge. Each member of a swarm acts by balancing between individual knowledge and group knowledge.

When solving problems using PSO, an objective function is formulated indicating the desired outcome. In this chapter, the objective function is the missing data estimation function represented by Eq. 7.6. To achieve an optimum missing data estimation function state, a social network representing a population of possible solutions is randomly generated. The individuals within this social network interrelate with their neighbours and are called particles. A process to update these particles is undertaken by assessing the fitness of each particle. Each particle is able to recall the position where it had its best success as measured by the missing data estimation function. The best solution of the particle is called the local best and each particle makes this information on the local best accessible to their neighbors and, in turn, also observe their neighbors' success.

The PSO was developed by Kennedy and Eberhart (1995) and it was inspired by algorithms that model the "flocking behavior" seen in birds. PSO has been very successful in optimizing complex problems. Marwala (2005) used PSO to improve finite element models to better reflect the measured data. This method was compared to a finite element model updating approach that used simulated annealing and a genetic algorithm. The proposed methods were tested on a simple beam and an unsymmetrical H-shaped structure. It was observed that, on average, the PSO method gave the most accurate results followed by simulated annealing and then the genetic algorithm.

Dindar and Marwala (2004) successfully used PSO to optimize the structure of a committee of neural networks. The results obtained from the optimized networks were found to give better results than both un-optimized networks and the committee of networks.

Ransome et al. (2005) successfully used PSO to optimize the position of a patient during radiation therapy. In this application, a patient positioning system integrating a robotic arm was designed for proton beam therapy. A treatment image was aligned with a pre-defined reference image and this was attained by aligning the radiation and reference field boundaries and then registering the patient's anatomy relative to the boundary. Methods for both field boundary and anatomy alignment, including particle swarm optimization, were implemented. It was found that the PSO was successful to overcome problems in existing solutions.

Farzi et al. (2013) applied PSO to choose the best portfolio in 50 supreme Tehran Stock Exchange companies and optimize the rate of return, risks, liquidity, and sharp ratio. The results were then compared to Markowitz's approach (Markowitz 1952) and genetic algorithms and it was observed that, although the return of the portfolio of PSO model was less than in Markowitz approach model, it was able to decrease the risk.

Nasir et al. (2012) applied a dynamic neighbourhood learning based particle swarm optimizer for global numerical optimization and the results indicated good performance on locating the global optimum solution on complicated and multi-modal fitness functions when compared to five other types of PSO. Muthukaruppan and Er (2012) applied the PSO for diagnosis of coronary artery disease while Kalatehjari et al. (2012) applied PSO for slope stability analysis of homogeneous soil slopes. Gholizadeh and Fattahi (2012) applied PSO for design optimization of tall steel buildings, while Karabulut and Ibrikci (2012) applied PSO to identify transcription factor binding sites.

When applying PSO, each particle which is represented by two vectors: $p_i(k)$ the position and $v_i(k)$ the velocity at step k. Positions and velocities of particles are randomly generated and then updated using the position of the best solution that a specific particle has encountered during the simulation called $pbest_i$ and the best particle in the swarm which is called $gbest(k)$. The updated velocity of a particle i can be estimated using the following equation (Kennedy and Eberhart 1995):

$$v_i(k + 1) = \gamma v_i(k) + c_1 r_1 (pbest_i - p_i(k)) + c_2 r_2 (gbest(k) - p_i(k)) \quad (7.8)$$

Here, γ is the inertia of the particle, c_1 and c_2 are the 'trust' parameters, r_1 and r_2 are random numbers between 0 and 1. In Eq. 7.8, the first expression is the current motion, the second expression is the particle memory influence, and the third expression is the swarm influence. The updated position of a particle i can be estimated using these equations (Kennedy and Eberhart 1995):

$$p_i(k + 1) = p_i(k) + v_i(k + 1) \quad (7.9)$$

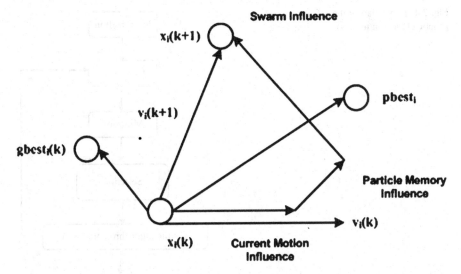

Fig. 7.3 Velocity and particle update in particle swarm optimization

The inertia of the particle regulates the relationship between the current velocity of the particle and the previous velocity. The trust parameter c_1 represents how much confidence the current particle has on itself, while the trust parameter c_2 represents the confidence the current particle has on the population. The parameters r_1 and r_2 are random numbers between 0 and 1 and they allow the swarm to explore the space.

The implementation of PSO can be summarized as follows, and is also shown in Fig. 7.3 (Kennedy and Eberhart 1995; Marwala 2010):

1. Randomly initialize a population of particles' positions and velocities.
2. Estimate the velocity for each particle in the swarm using Eq. 7.8.
3. Update the position of each particle using Eq. 7.9.
4. Repeat Steps 2 and 3 until convergence.

7.5 Genetic Algorithms (GA)

The missing data estimation method presented in this chapter also uses a genetic algorithm to estimate the missing data by minimizing Eq. 7.6. A genetic algorithm is a population-based, probabilistic technique that operates to identify a solution to a problem from a population of possible solutions (Goldberg 1989, 2002; Holland 1975; Marwala 2009). It is applied to identify estimated solutions to challenging problems through the similarity of the principles of evolutionary biology to computer science (Goldberg 2002; Marwala 2009; Tettey and Marwala 2006).

Fig. 7.4 Flow chart of the
genetic algorithm method

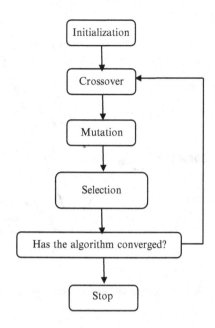

Fig. 7.4 Flow chart of the genetic algorithm method

It was derived from Darwin's theory of evolution where members of the population compete to survive and reproduce, while the weaker members die-out from the population.

Every individual has a fitness value indicating how well it fulfills the objective of solving the problem. New individual solutions are created during a cycle of generations, where selection and recombination operations occur, alike how gene transfer occurs to the current individuals. This continues until a termination condition is achieved, then the best individual by far is deemed to be the estimation for missing data. This chapter explains the application of a genetic algorithm to optimize Eq. 7.6.

Successful applications of the genetic algorithm include optimizing rough set partitions (Crossingham and Marwala 2008), missing data imputation (Hlalele et al. 2009), finite element updating (Marwala 2002, 2010), controlling fermentation (Marwala 2004), fault diagnosis (Marwala and Chakraverty 2006), HIV prediction (Leke et al. 2006), training neural networks (Marwala 2007), stock market prediction (Marwala et al. 2001), bearing fault classification (Mohamed et al. 2006), optimal weight classifier selection (Hulley and Marwala 2007) and call performance classification (Patel and Marwala 2009).

When applying the genetic algorithm, the following steps are followed: initialization, crossover, mutation, selection, reproduction, and termination. The three most important aspects of using a genetic algorithm are the definition of the objective function, implementation of the genetic representation, and implementation of the genetic operators (Marwala 2012). The details of genetic algorithms are shown in Fig. 7.4.

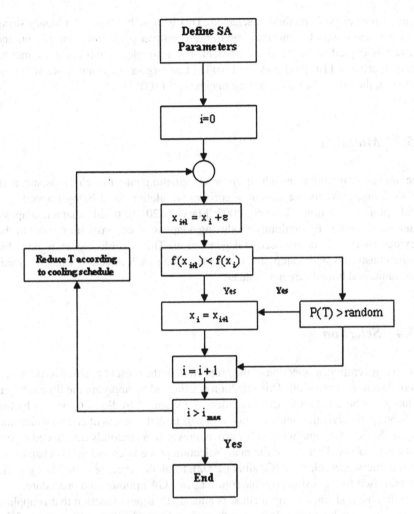

Fig. 7.5 The diagram of simulated annealing

7.5.1   *Initialization*

At this stage, a population of individual solutions is randomly created. This initial population is sampled so as to cover a good representation of the solution space.

7.5.2   *Crossover*

The crossover operator mixes genetic information in the population by cutting pairs of chromosomes at random points along their length and exchanging the cut sections over (Goldberg 2002, 1989; Marwala 2010; Banzhaf et al. 1998). In this chapter,

a one crossover point method is selected. This is done by copying a binary string
from the beginning of a chromosome to the crossover point from one parent, and
the rest is copied from the second parent. For example, if two chromosomes in
binary space $a = \mathbf{1100}1011$ and $b = 1101\mathbf{1111}$ undergo a one-point crossover at the
midpoint, then the resulting offspring may be $c = 11001111$.

7.5.3 Mutation

The mutation operator introduces new information into the chromosome and,
by so doing, prevents the genetic algorithm simulation from being trapped in a
local optimum solution (Goldberg 2002; Marwala 2010). In this chapter, adaptive
mutation is applied by randomly producing adaptive directions with respect to the
previous successful or unsuccessful generation. The feasible region is bounded
by the constraints and a step size is selected along each direction whereby linear
constraints and bounds are not violated.

7.5.4 Selection

In every generation, a selection of the proportion of the present population is chosen
to create a new population. This selection is achieved by applying the fitness-based
technique, where solutions that are fitter, as measured by Eq. 7.6, have a higher
probability of survival. Some selection methods rank the fitness of each solution and
choose the best solution, while other procedures rank a randomly designated aspect
of the population. There are quite many selection procedures and in this chapter we
use roulette-wheel selection (Goldberg 2002). Roulette-wheel selection is a genetic
operator used for choosing possible solutions in a GA optimization procedure.

In this method, each likely method is allocated a fitness function that is applied
to map the probability of selection with each individual solution. Let's say, if the
fitness f_i is of individual i in the population, then the probability that this individual
is chosen is (Goldberg 2002):

$$p_i = \frac{f_i}{\sum_{j=1}^{N} f_j} \qquad (7.10)$$

Here, N is the total population size.

This technique ensures that solutions with higher fitness values have higher
probabilities of survival than those with a lower fitness value. The benefit of this
is that, even though a solution may have a low fitness value, it may still have some
aspects that are advantageous in the future.

7.5.5 *Termination*

The technique described is repeated until a termination condition has been achieved, either for the reason that a chosen solution that satisfies the objective function has been identified or for the reason that a stated number of generations have been realized or the solution has converged or any combination of these.

7.6 Simulated Annealing (SA)

Simulated Annealing (SA) is a Monte Carlo technique that is applied to identify an optimal solution. It was inspired by the annealing process where metals re-crystalize or liquids freeze. In the annealing process, the object is heated until it is molten, then it is gradually cooled in such a way that the metal, at any time, is nearly in thermodynamic equilibrium. As the temperature of the object is cooled, the system becomes more organized and tends to a frozen state at $T = 0$. If the cooling procedure is done unsatisfactorily or the initial temperature of the object is not adequately high, the system may turn into a meta-stable state demonstrating that the system is stuck in a local minimum energy state.

   Liu et al. (2012) successfully applied the simulated annealing method in multi-criteria network path problems, while Shao and Zuo (2012) used it for higher dimensional projection depth. Milenkovic et al. (2012) successfully applied a fuzzy simulated annealing method for project time–cost trade-off, while Fonseca et al. (2012) applied simulated annealing to the high school timetabling problem. Other successful applications of simulated annealing include antenna array design in multi-input-multi-output radar (Dong et al. 2012), efficient bitstream extraction for scalable video (Wan et al. 2012), image reconstruction (Martins et al. 2012) and optimal sensor placement (Tian et al. 2012).

   Simulated annealing has its origins from the work of Metropolis et al. (1953) and it comprises selecting the initial state and temperature, maintaining temperature constant, changing the initial formation, and calculating the error at the new state. If the new error is lower than the old error, then accept the new state, otherwise if the error is higher, then accept this state with a low probability. Simulated annealing substitutes a current solution with a "nearby" random solution with a probability that depends on the difference between the corresponding function values and the temperature. The temperature drops during the course of the procedure until it approaches zero and at this stage there are less random changes in the solution. Simulated annealing identifies the global optimum but it can reach infinite time in doing so. The probability of accepting the reversal is given by Boltzmann's equation (Černý 1985):

$$P(\Delta E) = \frac{1}{Z} \exp\left(-\frac{\Delta E}{T}\right) \tag{7.11}$$

Here, ΔE is the variance in error between the old and new states, T is the temperature of the system and Z is the normalization factor that guarantees that when the probability is integrated over to infinity it becomes 1.

7.6.1 Simulated Annealing Parameters

As described by Marwala (2010), applying simulated annealing means that a number of parameters and selections require to be stated: the state space, the objective function, the candidate generator process, the acceptance probability function, and the annealing temperature schedule. The selection of these parameters is important because it has an impact on the efficacy of the SA technique. Nevertheless, there is no optimal mode for selecting these parameters that will function for all problems and there is also no methodical routine of optimally selecting these parameters for a given problem. Accordingly, the selection of these parameters is mainly subjective and the technique of trial and error is extensively applied.

7.6.2 Transition Probabilities

When SA is applied, a random walk procedure is used for a given temperature. This random walk procedure involves moving from one temperature to another. The probability of moving from one state to another is called the transition probability. This probability is dependent on the current temperature, the order of producing the candidate solution, and the acceptance probability function. In this chapter, a Markov Monte Carlo (MMC) technique is applied to ensure a transition from one state to another. The MMC generates a chain of possible missing data estimates and accepts or rejects them using the Metropolis algorithm (Metropolis et al. 1953).

7.6.3 Monte Carlo Method

The Monte Carlo technique is a computational procedure that applies recurring random sampling to estimate a result (Arya et al. 2012; Klopfer et al. 2012). Jeremiah et al. (2012) applied Monte Carlo sampling for efficient hydrological model parameter optimization, while Giraleas et al. (2012) applied the Monte Carlo procedure for analysing productivity change using growth accounting and frontier-based approaches. Fang et al. (2012) applied Monte Carlo simulation for variability quantification in finite element models. Other applications of Monte Carlo simulation include evaluating reliability indices accounting omission of random repair time for distribution systems (Arya et al. 2012) and characterization and optimization of pyroelectric X-ray sources (Klopfer et al. 2012).

7.6.4 Markov Chain Monte Carlo (MCMC)

MCMC is a procedure of simulating a chain of states through a random walk. It entails a Markov process and a Monte Carlo simulation (Sheridan et al. 2012). Fishman (2012) successfully applied MCMC for counting contingency tables, while Hettiarachchi et al. (2012) successfully applied a marginalized Markov Chain Monte Carlo method for model based analysis of EEG data. Botlani-Esfahani and Toroghinejad (2012) successfully applied a Bayesian neural network and the reversible jump Markov Chain Monte Carlo Method to forecast the grain size of hot strip low carbon steels while Laloy et al. (2012) successfully applied the MCMC to analyse mass conservative three-dimensional water tracer distribution. Stošić et al. (2012) successfully applied the MCMC to optimize river discharge measurements.

If a system whose evolution is expressed by a stochastic process $\{x_1, x_2, \ldots, x_i\}$ of random variables is considered, a random variable x_i inhabits a state x at discrete time i. The list of all states that all random variables can probably occupy is known as the state space. If the probability that the system is in state x_{i+1} at time $i+1$ depends entirely on the fact that it was in state x_i at time i, then the random variables $\{x_1, x_2, \ldots, x_i\}$ form a Markov chain. For MCMC, the transition between states is attained by introducing a random noise (ε) to the current state as follows (Laloy et al. 2012):

$$x_{i+1} = x_i + \varepsilon \tag{7.12}$$

7.6.5 Acceptance Probability Function: Metropolis Algorithm

When the present state has been attained, it is either accepted or rejected. In this chapter, the acceptance of a state is conducted using the Metropolis algorithm (Metropolis et al. 1953; Shao et al. 2012; Vihola 2012; Lee et al. 2012). Zhou et al. (2012) successfully applied Metropolis-Hastings sampling for system error registration. In the Metropolis procedure, on sampling a stochastic process $\{x_1, x_2, \ldots, x_n\}$ consisting of random variables, random changes to x are introduced and are either accepted or rejected according to the following criterion:

$$if \ E_{new} < E_{old} \ accept \ state \ (s_{new})$$

$$else$$

$$accept \ (s_{new}) \ with \ probability$$

$$\exp\{-(E_{new} - E_{old})\} \tag{7.13}$$

Here, E is the objective function.

7.6.6   Cooling Schedule

Cooling scheduling is the procedure which is followed to lower the temperature T (Stander and Silverman 1994). Natural annealing teaches us that the cooling rate should be adequately low for the probability distribution of the present state to be close to the thermodynamic equilibrium at all times during the simulation (Miki et al. 2003). The time taken for the equilibrium to be restored after a change in temperature is influenced by the shape of the objective function, the current temperature and the candidate generator. The best cooling rate should be experimentally attained for each problem. Thermodynamic, simulated annealing circumvents this problem by removing the cooling schedule and regulating the temperature at each step in the simulation based on the difference in energy between the two states, in accordance to the laws of thermodynamics (Weinberger 1990). The following cooling model is used (Salazar and Toral 1997; Marwala 2010):

$$T(i) = \frac{T(i - 1)}{1 + \sigma} \tag{7.14}$$

where $T(i)$ is the current temperature; $T(i - 1)$ is the previous temperature and σ is the cooling rate. The implementation of SA is shown in Fig. 7.1 (Marwala 2010).

7.7   Experimental Investigations and Results

The methodology described below is used to analyze the manufacturing data from South Africa collected between 1992 and 2011. The variables identified for the modelling are (1) Domestic sales volumes; (2) Production volumes; (3) Number of factory workers; (4) Current stocks of raw materials in relation to planned production; (5) Business confidence; (6) Percentage rating shortage of skilled labour a constraint; and (7) Percentage rating shortage of semi-skilled labour a constraint. We then build an auto-associative network with seven input variables, four hidden nodes and seven outputs. The auto-associative network was based on the multi-layer perceptron architecture. It had a hyperbolic tangent activation function in the hidden nodes and a linear activation function in the outer layer. It assumes that the production volumes will be treated as a missing values to be estimated. The missing data estimation equation is optimized using a genetic algorithm, particle swarm optimization, and simulated annealing to identify the production volume.

When implementing a genetic algorithm, the population size was set to 20, the number of generations was set to 100, and the one point crossover with probability of crossover was set to be 0.65. The selection function which was used was the Roulette wheel and adaptive mutation with a mutation rate of 0.01. On implementing simulated annealing, the initial temperature was set to be 100. When implementing PSO, a population size of ten was set and the simulation was conducted for 500 generations. The results obtained when these optimization methods were implemented were an average of 5.3 % for GA, 5.1 % for simulated annealing, and 5.4 % for the PSO.

7.8 Conclusion

This chapter introduced auto-associative networks with genetic algorithms, particle swarm and simulated annealing optimization methods for modelling manufacturing data. The autoassociative network was created using the multi-layered perceptron. The results obtained gave marginally best results for simulated annealing, followed by genetic algorithm and the particle swarm optimization method.

References

Abdella M (2005) The use of genetic algorithms and neural networks to approximate missing data in database. Master's thesis, University of the Witwatersrand, Johannesburg

Abdella M, Marwala T (2005) Treatment of missing data using neural networks. In: Proceedings of the IEEE international joint conference on neural networks, Montreal, 2005, pp 598–603

Abdella M, Marwala T (2006) The use of genetic algorithms and neural networks to approximate missing data in database. Comput Inf 24:1001–1013

Arya LD, Choube SC, Arya R, Tiwary A (2012) Evaluation of reliability indices accounting omission of random repair time for distribution systems using Monte Carlo simulation. Int J Electr Power Energy Syst 42:533–541

Banzhaf W, Nordin P, Keller R, Francone F (1998) Genetic programming-an introduction: on the automatic evolution of computer programs and its applications. Morgan Kaufmann, San Francisco

Botlani-Esfahani M, Toroghinejad MR (2012) Application of a Bayesian artificial neural network and the reversible jump Markov chain Monte Carlo method to predict the grain size of hot strip low carbon steels. J Serb Chem Soc 77:937–944

Černý V (1985) Thermodynamical approach to the traveling salesman problem: an efficient simulation algorithm. J Optim Theory Appl 45:41–51

Crossingham B, Marwala T (2008) Using genetic algorithms to optimise rough set partition sizes for HIV data analysis. Adv Intell Distrib Comput Stud Comput Intell 78:245–250

Dervilis N, Barthorpe R, Antoniadou I, Staszewski WJ, Worden K (2012) Damage detection in carbon composite material typical of wind turbine blades using auto-associative neural networks. In: Proceedings of SPIE – the international society for optical engineering, San Diego, 2012, art. no. 834806

Dindar ZA, Marwala T (2004) Option pricing using a committee of neural networks. Proc IEEE Int Conf Syst Man Cybern 1:434–438

Dong J, Yang J, Lei W, Shi R, Guo Y (2012) Antenna array design in MIMO radar using cyclic difference sets and simulated annealing. In: Proceedings of the international conference on microwave and millimeter wave technology, Shenzhen, China, pp 237–240

Engebrecht AP (2005) Fundamentals of computational swarm intelligence. Wiley, New York

Fang S-E, Ren W-X, Perera R (2012) A stochastic model updating method for parameter variability quantification based on response surface models and Monte Carlo simulation. Mech Syst Signal Process 33:83–96

Farzi S, Shavazi AR, Pandari AR (2013) Using quantum-behaved particle swarm optimization for portfolio selection problem. Int Arab J Inf Technol 10:art. no. 2/7-2761

Fishman GS (2012) Counting contingency tables via multistage Markov chain Monte Carlo. J Comput Graph Stat 21:713–738

Fonseca GHG, Brito SS, Santos HG (2012) A simulated annealing based approach to the high school timetabling problem. Lect Note Comput Sci 7435:540–549

Gholizadeh S, Fattahi F (2012) Design optimization of tall steel buildings by a modified particle swarm algorithm. Struct Design Tall Spec Build. doi:10.1002/tal.1042

Giraleas D, Emrouznejad A, Thanassoulis E (2012) Productivity change using growth accounting and frontier-based approaches – evidence from a Monte Carlo analysis. Eur J Oper Res 222:673–683

Goldberg DE (1989) Genetic algorithms in search, optimization, and machine learning. Addison-Wesley, Reading

Goldberg DE (2002) The design of innovation: lessons from and for competent genetic algorithms. Addison-Wesley, Reading

Granville V, Krivanek M, Rasson J-P (1994) Simulated annealing: a proof of convergence. IEEE Trans Pattern Anal Mach Intell 16:652–656

Hettiarachchi I, Mohamed S, Nahavandi S (2012) A marginalised Markov chain Monte Carlo approach for model based analysis of EEG data. In: Proceedings of the international symposium on biomedical imaging, Barcelona, 2012, pp 1539–1542

Hlalele N, Nelwamondo FV, Marwala T (2009) Imputation of missing data using PCA, neuro-fuzzy and genetic algorithms. Lect Note Comput Sci 5507:485–492

Holland JH (1975) Adaptation in natural and artificial systems. University of Michigan Press, Ann Arbor

Hulley G, Marwala T (2007) Genetic algorithm based incremental learning for optimal weight and classifier selection. Comput Model Life Sci Am Inst Phys Ser 952:258–267

Jeremiah E, Sisson SA, Sharma A, Marshall L (2012) Efficient hydrological model parameter optimization with sequential Monte Carlo sampling. Environ Model Softw 38:283–295

Kalatehjari R, Ali N, Hajihassani M, Kholghi Fard M (2012) The application of particle swarm optimization in slope stability analysis of homogeneous soil slopes. Int Rev Model Simul 5:458–465

Karabulut M, Ibrikci T (2012) A Bayesian scoring scheme based particle swarm optimization algorithm to identify transcription factor binding sites. Appl Soft Comput J 12:2846–2855

Kennedy J (1997) The particle swarm: social adaptation of knowledge. In: Proceedings of IEEE international conference on evolutionary computation, Piscataway, 1997, pp 303–308

Kennedy J, Eberhart R (1995) Particle swarm optimization. In: Proceedings of IEEE international conference on neural networks, Piscataway, 1995, pp 1942–1948

Kennedy J, Eberhart RC (2001) Swarm intelligence. Morgan Kaufmann, San Francisco

Kim JK, Shin DW (2012) The factoring likelihood method for non-monotone missing data. J Korean Stat Soc 41:375–386

Kirkpatrick S, Gelatt CD, Vecchi MP (1983) Optimization by simulated annealing. Science 220:671–680

Klopfer M, Wolowiec T, Satchouk V, Alivov Y, Molloi S (2012) Characterization and optimization of pyroelectric X-ray sources using Monte Carlo spectral models. Nucl Instrum Method Phys Res Sec A Accel Spectrom Detect Assoc Equip 689:47–51

Kramer MA (1992) Autoassociative neural networks. Comput Chem Eng 16:313–328

Laloy E, Linde N, Vrugt JA (2012) Mass conservative three-dimensional water tracer distribution from MCMC inversion of time-lapse GPR data. Water Resour Res. doi:10.1029/2011WR011238 (in press)

Lee C-H, Xu X, Eun DY (2012) Beyond random walk and metropolis-hastings samplers: why you should not backtrack for unbiased graph sampling. Perform Eval Rev 40:319–330

Leke B, Marwala T, Tim T, Lagazio M (2006) Using genetic algorithms versus line search optimization for HIV predictions. Trans Inf Sci Appl 4:684–690

Lemma TA, Hashim FM (2012) Wavelet analysis and auto-associative neural network based fault detection and diagnosis in an industrial gas turbine. In: Proceedings of the IEEE business, engineering and industrial applications colloquium, Kuala Lumpur, 2012, pp 103–108

Liu L, Mu H, Luo H, Li X (2012) A simulated annealing for multi-criteria network path problems. Comput Oper Res 39:3119–3135

Makki B, Hosseini MN (2012) Some refinements of the standard autoassociative neural network. Neural Comput Appl. doi:10.1007/s00521-012-0825-5 (in press)

Markowitz H (1952) Portfolio selection. J Finance 7:77–91

Martins TDC, De Camargo EDLB, Lima RG, Amato MBP, Tsuzuki MDSG (2012) Image reconstruction using interval simulated annealing in electrical impedance tomography. IEEE Trans Biomed Eng 59:1861–1870

Marwala T (2002) Finite element updating using wavelet data and genetic algorithm. Am Inst Aeronaut Astronaut J Aircraft 39:709–711

Marwala T (2004) Control of complex systems using Bayesian neural networks and genetic algorithm. Int J Eng Simul 5:28–37

Marwala T (2005) Finite element model updating using particle swarm optimization. Int J Eng Simul 6:25–30

Marwala T (2007) Bayesian training of neural network using genetic programming. Pattern Recognit Lett 28:452–1458

Marwala T (2009) Computational intelligence for missing data imputation, estimation and management: knowledge optimization techniques. IGI Global Publications, New York

Marwala T (2010) Finite element model updating using computational intelligence techniques. Springer, London

Marwala T (2012) Condition monitoring using computational intelligence methods. Springer, London

Marwala T, Chakraverty S (2006) Fault classification in structures with incomplete measured data using autoassociative neural networks and genetic algorithm. Curr Sci 90:542–548

Marwala T, Lagazio M (2011) Militarized conflict modeling using computational intelligence techniques. Springer, London

Marwala T, de Wilde P, Correia L, Mariano P, Ribeiro R, Abramov V, Szirbik N, Goossenaerts J (2001) Scalability and optimisation of a committee of agents using genetic algorithm. In: Proceedings of the international symposium on soft computing and intelligent systems for industry, Florida, USA, arXiv 0705.1757

Metropolis N, Rosenbluth AW, Rosenbluth MN, Teller AH, Teller E (1953) Equation of state calculations by fast computing machines. J Chem Phys 21:1087

Miki M, Hiroyasu T, Wako J, Yoshida T (2003) Adaptive temperature schedule determined by genetic algorithm for parallel simulated annealing. In: Proceedings of the congress on evolutionary computation, Canberra, Australia, pp 459–466

Milenkovic M, Bojovic N, Ribeiro RA, Glisovic N (2012) A fuzzy simulated annealing approach for project time–cost tradeoff. J Intell Fuzzy Syst 23:203–215

Miranda V, Castro ARG, Lima S (2012) Diagnosing faults in power transformers with autoassociative neural networks and mean shift. IEEE Trans Power Deliv 27:1350–1357

Mohamed S, Tettey T, Marwala T (2006) An extension neural network and genetic algorithm for bearing fault classification. In: Proceedings of the IEEE international joint conference on neural networks, Vancouver, Canada, pp 7673–7679

Moreau V, Bage G, Marcotte D, Samson R (2012) Statistical estimation of missing data in life cycle inventory: an application to hydroelectric power plants. J Clean Prod 37:335–341

Muthukaruppan S, Er MJ (2012) A hybrid particle swarm optimization based fuzzy expert system for the diagnosis of coronary artery disease. Expert Syst Appl 39:11657–11665

Nasir M, Das S, Maity D, Sengupta S, Halder U, Suganthan PN (2012) A dynamic neighborhood learning based particle swarm optimizer for global numerical optimization. Inf Sci 209:16–36

Nelwamondo FV (2008) Computational intelligence techniques for missing data imputation. Ph.D. thesis, University of the Witwatersrand, Johannesburg

Nelwamondo FV, Marwala T (2007) Rough set theory for the treatment of incomplete data. In: Proceedings of the IEEE conference on fuzzy systems, London, UK, pp 338–343

Palmé T, Breuhaus P, Assadi M, Klein A, Kim M (2011) Early warning of gas turbine failure by nonlinear feature extraction using an auto-associative neural network approach. Proc ASME Turbo Expo 3:293–304

Patel PB, Marwala T (2009) Genetic algorithms, neural networks, fuzzy inference system, support vector machines for call performance classification. In: Proceedings of the IEEE internationall conference on machine learning and applications, Florida, USA, pp 415–420

Ransome TM, Rubin DM, Marwala T, de Kok EA (2005) Optimising the verification of patient positioning in proton beam therapy. In: Proceedings of the IEEE 3rd international conference on computational cybernetics, Mauritius, 2005, pp 279–284

Rey-del-Castillo P, Cardeñosa J (2012) Fuzzy min-max neural networks for categorical data: application to missing data imputation. Neural Comput Appl 21:1349–1362

Salazar R, Toral R (1997) Simulated annealing using hybrid Monte Carlo. J Stat Phys 89(5/6):1047–1060

Shao W, Zuo Y (2012) Simulated annealing for higher dimensional projection depth. Comput Stat Data Anal 56:4026–4036

Shao W, Guo G, Meng F, Jia S (2012) An efficient proposal distribution for metropolis-hastings using a B-splines technique. Comput Stat Data Anal 57:465–478

Sheridan P, Yagahara Y, Shimodaira H (2012) Measuring preferential attachment in growing networks with missing-timelines using Markov chain Monte Carlo. Physica A Stat Mech Appl 391:5031–5040

Shi Y, Eberhart RC (1998) A modified particle swarm optimizer. In: Proceedings of IEEE international conference on evolutionary computation, Anchorage, Alaska, pp 69–73

Sivaram GSVS, Ganapathy S, Hermansky H (2010) Sparse auto-associative neural networks: theory and application to speech recognition. In: Proceedings of the 11th annual conference of the international speech communication association, Florence, Italy, pp 2270–2273

Stander J, Silverman BW (1994) Temperature schedules for simulated annealing. Stat Comput 4:21–32

Stošić BD, Santos Silva JR, Filho MC, Barros Cantalice JR (2012) Optimizing river discharge measurements using Monte Carlo Markov chain. J Hydrol 450–451:199–205

Tettey T, Marwala T (2006) Controlling interstate conflict using neuro-fuzzy modeling and genetic algorithms. In: Proceedings of the 10th IEEE international conference on intelligence engineering systems, London, UK, pp 30–44

Tian L, Chen H-G, Zhu J, Zhang L-S, Chen W-H (2012) A study of optimal sensor placement based on the improved adaptive simulated annealing genetic algorithms. J Vib Eng 25:238–243

Tsai LT, Yang C-C (2012) Improving measurement invariance assessments in survey research with missing data by novel artificial neural networks. Expert Syst Appl 39:10456–10464

Vihola M (2012) Robust adaptive Metropolis algorithm with coerced acceptance rate. Stat Comput 22:997–1008

Wan S, Yang K, Zhou H (2012) Efficient bitstream extraction for scalable video based on simulated annealing. Concurr Comput Pract Exp 24:1223–1230

Wang C, Yang Y (2011) Robust face recognition from single training image per person via auto-associative memory neural network. In: Proceedings of the international conference on electrical and control engineering, Beijing, China, pp 4947–4950

Weinberger E (1990) Correlated and uncorrelated fitness landscapes and how to tell the difference. Biol Cybern 63:325–336

Zhou L, Liang Y, Pan Q (2012) System error registration based on Metropolis-Hastings sampling. Syst Eng Electron 34:433–438

Chapter 8
Correlations Versus Causality Approaches to Economic Modeling

Abstract This chapter explores the issue of treating a predictive system as a missing data problem i.e. correlation exercise and compares it to treating as a cause and effect exercise, that is, feed-forward network. An auto-associative neural network is combined with genetic algorithm and then applied to missing economic data estimation. The algorithm is used on data that contain ten economic variables. The results of the missing data imputation approach are compared to those from a feed-forward neural network.

8.1 Introduction

This chapter explores the problem of correlation and causal approaches to modeling economic data. The concept of correlation is sometimes confused with the concept of correlation. It turns out that if variable x is causes variable y to happen then there is necessarily a correlation between variable x and variable y. What will make the correlation between variable x and variable y to necessarily imply causality is that variable x should happen before variable y. This time delay necessarily implies that once x had happened then y happen.

To build the causal machine we create a model which takes variables x as inputs and give output variable y. There are many models that have been used to achieve this goal and these include linear as well as non-linear models. In this chapter we model economic data as well as credit scoring data and these are complicated data sets. Consequently, it is important to use non-linear models to model these data sets and in this chapter we apply the multi-layered perceptron neural networks.

To build a correlation machine we take variables x and y as a combined variable z and map this z variable non-linearly on itself. This is called the autoassociative neural networks. The auto-associative model basically models the interrelationships between variables and thus we call this approach a correlation machine. For regression or classification the dependent variable y is estimated using an optimization

T. Marwala, *Economic Modeling Using Artificial Intelligence Methods*, Advanced
Information and Knowledge Processing, DOI 10.1007/978-1-4471-5010-7_8,
© Springer-Verlag London 2013

method or for a two class problem by testing each class and evaluating whether the interrelationships that have been defined by the auto-associative network are maintained as much as possible.

The other issue that is explored in this chapter is the Granger causality. We apply the automatic relevance determination (ARD) method to evaluate the causal relationships amongst variables. In the ARD implementation we define the concept of relevance of a variable in as far as the prediction of the dependent variable as another way of articulating causality.

8.2   Causality Approach to Economic Modeling

In this chapter we also apply the input–output to model economic data. If we have an input vector x and output vector y where the output happens after a time lag t after x has happened and this can be written as follows:

$$y = f(x, w) \tag{8.1}$$

Here w is the free parameter also known as weights that ensure that whenever x is given then y can be estimated. This relationship necessarily assumes that there is a causal relationship between x and y.

In this chapter this relationship is represented using a multi-layered percep-tron (MLP) neural networks which has been described in detail in the previous chapters. This relationship is written mathematically as follows (Marwala 2012; Bishop 1995):

$$y_k = f_{outer}\left(\sum_{j=0}^{M} w_{kj}^{(2)} f_{inner}\left(\sum_{i=0}^{d} w_{ji}^{(1)} x_i\right)\right) \tag{8.2}$$

Here y_k is the kth output values, f_{outer} is the activation function in the outer layer and in this chapter it is a linear function while f_{inner} is the activation function in the inner layer and in this chapter it is a hyperbolic tangent function, w is the network weight and d is the size of input vector and M is the number of the hidden units. The diagram illustrating a model in Eq. 8.2 is shown in Fig. 8.1.

Turchenko et al. (2011) applied the MLP for stock price prediction and the results showed high accuracy of the prediction. Correa and Gonzalez (2011) applied the MLP for credit scoring case and the results demonstrated that the MLP performed better than the logistic regression. Ebrahimpour et al. (2011) applied the MLP to model the Tehran Stock Exchange whereas Sermpinis et al. (2012) successfully applied the MLP to predict the Euro-US-Dollar exchange rate.

Lin (2012) successfully applied the MLP to design power system transient stability preventive control system whereas Daya et al. (2012) successfully applied the MLP for multiclass vehicle type recognition. Pereira et al. (2012) successfully

Fig. 8.1 A diagram showing
a multi-layer perceptron
neural network

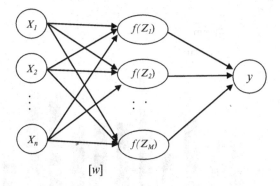

applied the MLP for fingerprint detection, while Salazar et al. (2012) implemented
data transformations and seasonality adjustments to improve the MLP ensembles.
Other successful applications of the MLP include blind modulation classification
(Dubey et al. 2012), drought forecasting (Rezaeian-Zadeh and Tabari 2012),
electricity load forecasting (Dragomir et al. 2011), automotive price prediction
(Peyghami and Khanduzi 2012) and landmine detection (Achkar et al. 2011).

In this section, we implement the MLP to predict the consumer price index (CPI)
as it was done in Chap. 3. The input variables to the neural networks is the mining
production, transport, storage and communication, financial intermediation, insurance, real estate and business services, community, social and personal services,
gross value added at basic prices, taxes less subsidies on products, affordability,
economic growth, repo rate, gross domestic product, household consumption and
investment from the South African economy. The neural network was trained with
12 input units, 23 hidden nodes and one output nodes. The prediction accuracy
realized was 84.58 % and the sample results are shown in Fig. 8.2.

8.3   Correlation Machines for Economic Modeling

In the previous section we deemed the MLP input–output model to be a causal
model. This is because this model fundamentally assumes that there is a relationship
between the input (causes) and output (effect). In this section we describe another
model which is largely based on the interrelationships amongst variables and this
model is called the correlation machine. It is based on the autoassociative neural
network which essentially maps a set of variables (this would include both the input
and the output in the previous model) to itself (Kramer 1992). This model is called
an autoassociative network and can be found in Fig. 8.3. Once this model is trained it
essentially defines correlation relationships between the data. Thereafter, this model
is treated as a missing data estimator where the missing data (the effect) is treated
as a variable to be estimated using an optimization method.

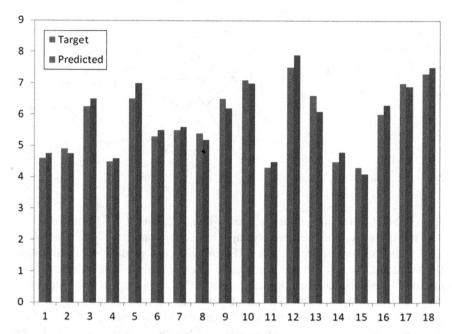

Fig. 8.2 Figure showing the results achieved using the MLP network

Fig. 8.3 An autoassociative model

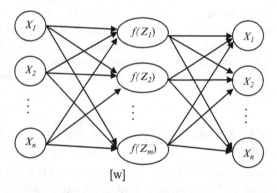

Cette and de Jong (2013) studied the breakeven inflation rates and their correlation relationships whereas Vladislavleva et al. (2012) predicted the energy output of wind farms based on weather data correlations.

Chen (2010) applied autoassociative model and genetic algorithm to estimate business failure. They conducted this by employing a method proposed by Abdella and Marwala (2005a, b) which solves a regression problem by treating the required variables as missing values to be estimated using the autoassociative neural networks and genetic algorithm. Further work on this subject can be found

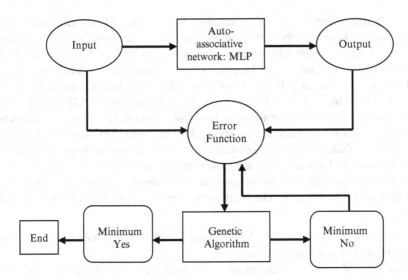

Fig. 8.4 The correlation machine

in Marwala (2009), Marwala and Chakraverty (2006) and Mistry et al. (2009). This procedure was observed to be feasible for predicting business failure by treating it as a missing value.

Aydilek and Arslan (2012) introduced a hybrid approach to estimate missing values in databases. The difference between this method and that explained in Marwala (2009) is that they used a hybrid neural network instead of the standard MLP and that they used K-nearest neighbor to enhance the optimization process. Narayanan et al. (2002) successfully applied auto-associative regression machines for missing data estimation.

The correlation machine proposed in this chapter is shown in Fig. 8.4. The implementation of the correlation machine in this chapter is as follows:

1. Create an autoassociative network using the MLP network
2. Formulate the missing data estimator and this is shown in Eq. 8.3
3. Minimize the missing data estimator using genetic algorithm
4. Stop when the method has converged.

The missing data estimator is written as follows (Marwala 2009):

$$E = \left\| \left(\left\{ \begin{matrix} \{X_k\} \\ \{X_u\} \end{matrix} \right\} - f \left(\left\{ \begin{matrix} \{X_k\} \\ \{X_u\} \end{matrix} \right\}, [W] \right) \right) \right\| \tag{8.3}$$

Here, E is the fitness function while the subscript u stands for the unknown component of X while the subscript k stands for known component of X.

In this chapter the optimization method chosen to solve the missing data optimization is the genetic algorithm (GA). Genetic algorithm is a mathemmatical procedure that optimizes a function based on the principles of evolution. In essence the process seeks to identify a set of parameters that are the fittest. Within the context of this chapter this essentially means identifying X_u in Eq. 8.3. Mutlu et al. (2013) applied a genetic algorithm technique for the assembly line worker assignment and balancing problem, whereas Vidal et al. (2013) applied a genetic algorithm for vehicle routing. Ramírez Palencia and Mejía Delgadillo (2012) applied a genetic algorithm for a bus body assembly line while Liu et al. (2012) applied a genetic algoritm in aviation ammunition transport. Other successful applications of genetic algorithms include customer segmentation (Davis 2012), dampening an impact of price fluctuation (Lu et al. 2012) and in scheduling problems (Sioud et al. 2012).

A genetic algorithm operates by using three key drivers, and these are crossover, mutation and reproduction. The implementation of genetic algorithms is as follows:

1. Initialize the algorithm by choosing the population size and generating the population of possible solutions
2. Pick pairs of possible solutions and cross-over them (this should be conducted with a certain probability)
3. Pick a set of possible solutions and mutate them (this should be conducted with a certain probability)
4. Reproduce the solution
5. Go to Step 2 and repeat until convergence or until the specified number of iterations have been reached

In this chapter we implement simple crossover and this is done iin binary space. Suppose we have two possible solutions which can be written in binary space as follows: 0101 and 1100. A cut-off point is randomly chosen and suppose in this chapter we choose after a second digit then the crossover is implemented as by exchanging the genes between the two possible solutions after this cut-off point. Thus 0101 and 1100 possibly becomes 0100 and 1101.

In this chapter we implement simple mutation. Again suppose we randomly choose a possible solution written in binary space as 0100 for mutation then we pick a gene for mutation randomly and if it is the second element of this possible solution then we invert this gene and thus 0100 becomes 0000.

Reproduction essentially means reproducing the solutions that are successful and in this chapter we apply the Roulette wheel reproduction. This is done by evaluating the fitnesses of all the solutions in the population and ranking them according to their fittness. The fittest solutions are reproduced with a higher probability than than the less fit ones.

The correlation machine was applied to estimate the CPI by treating it as a missing variable. The architecture of the autoassociative model was as follows: number of input and output were each 13, number of hidden nodes was ten, activation function was a hyperbolic tangent function in the inner layer and linear in the outer layer. The population size of the genetic algorithm was ten. When the method was implemented an accuracy of 88 % was observed and the sample results are in Fig. 8.5.

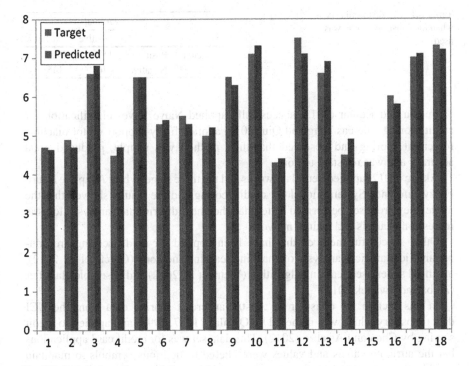

Fig. 8.5 Results from the correlation machine

The results obtained in this section and Sect. 8.2 seem to suggest that the correlation machine performs better than the causal machine. Of course this is specific to the problem at hand because there are other factors that might be at play, for example, the specific nature of the data and the optimization technique used. The next section compares the two procedures using the classification data.

8.4 Classification: Correlation and Causal Machine

In this section we apply the correlation and causal machines for credit scoring. Credit scoring is a problem where financiers assess individuals that are seeking credit as to whether they should be given credit or not and if they are given credit the terms of the credit granted. In this chapter we treat credit scoring as either a yes or a no credit.

Blanco et al. (2013) applied multilayer perceptron neural networks for credit scoring models for the micro-finance industry. They used data from a Peruvian micro-finance institution. The results obtained showed that that multi-layer perceptron neural network models gave good results.

Table 8.1 Confusion matrix
when the causal machine was
used

		Actual	
		Positive	Negative
Predicted	Positive	119	30
	Negative	46	145

Vedala and Kumar (2012) successfully applied Naive Bayes classification for credit scoring whereas Tang and Qiu (2012) applied fuzzy support vector machine for credit scoring and observed that their method was simple, produced good accuracy and was resistant to noise.

Akkoç (2012) applied neural networks and the three stage hybrid Adaptive Neuro Fuzzy Inference System model for credit scoring and the results showed that the model they proposed performed better than the linear discriminant analysis, logistic regression analysis and natural network.

Other successful methods that have been applied for credit scoring are non-parametric statistical analysis of machine learning methods (García et al. 2012), manifold supervised learning algorithm (Vieira et al. 2012) and ensemble classifier (Zieba and Światek 2012).

In this section we consider an Australian credit scoring data from the UCI data repository which was gathered by Quinlan (1987, 1992) and applied to credit scoring by Chen and Åstebro (2012). This data set was for credit card applications but the attribute names and values were altered to fictitious symbols to maintain confidentiality of the data. The data had a good combination of attributes which were continuous, nominal with small numbers of values and nominal with larger numbers of values. This data set had 14 input variables and the results was a yes or no in terms of the approval or disapproval of the credit card.

The causal and the correlation machines were created in order to classify the data. On creating the causal machine a multi-layer perceptron neural networks was used and it was trained using the scaled conjugate gradient method (Bishop 1995). The MLP network had the following attributes: number of inputs was 14, number of hidden units was eight, a number of output units was a one, activation functions in the hidden units was a hyperbolic tangent function while the activation function in the outer layer was a logistic function.

The network was trained with 350 data points and was tested with 340 data points. The results obtained was expressed in terms of the confusion matrix with the class cut-off at 0.5 and a 0 indicating no credit card be given while a 1 indicates that a credit card be given and the accuracy obtained was 78 % and confusion matrix is given in Table 8.1.

On creating the correlation machine a multi-layer perceptron autoassociative neural networks was used and it was trained using the scaled conjugate gradient method (Bishop 1995). The MLP network had the following attributes: number of inputs was 15 (here we include the outcome class), number of hidden units was eight, a number of output units was a 15, activation functions in the hidden units was a hyperbolic tangent function while the activation function in the outer layer was a logistic function. Instead of applying genetic algorithm in this case and because the

Table 8.2 Confusion matrix when the correlation machine was used

		Actual	
		Positive	Negative
Predicted	Positive	112	37
	Negative	58	133

answer was either a 0 for no credit and a 1 for a credit the correlation machine was obtained by testing a 0 or a 1 on the missing data estimator which is in Eq. 8.3 and the choosing the value which offers lower missing data estimator error.

The network was trained with 350 data points and was tested with 340 data points. The results obtained was expressed in terms of the confusion matrix with the class cut-off at 0.5 and a 0 indicating no credit card be given while a 1 indicates that a credit card be given and the accuracy obtained was 72 % and confusion matrix is given in Table 8.2.

The results obtained indicated that the causal machine performed better than the correlation machine.

8.5 Causality

Causality is when an event causes another event. This effect is, therefore, a result of the cause (Abdoullaev 2000; Pearl 2000; Green 2003). In engineering, a causal system is a structure with output and internal states that are caused only by the existing and preceding input values. In economics previous data are used to surmise causality by regression approaches (Granger 1969; Hacker and Hatemi 2006).

Miyazaki and Hamori (2013) used non-uniform weighting cross-correlations to test causality between gold return and the performance of the stock market. They observed one-direction causality in mean from stock to gold, however, they identified no causality in variance between the two. For the period prior to the financial crisis they observed two-direction causality and one-direction causality in mean and variance from stock to gold for the crisis period.

Liu and Wan (2012) applied cross-correlation analysis, structural co-integration and non-linear causality test to study the relationships between Shanghai stock market and CNY/USD exchange rate. They then applied the linear and non-linear Granger causality tests and observed no causality between stock prices and exchange rates.

Chen and Yeh (2012) studied the causality between demand uncertainty and hotel failure. This analysis was based on two phases with the first using a first-order autoregressive model to model lodging demand uncertainty and the second using a logit model to approximate the probability of failures of hotels. They observed that demand uncertainty caused hotel failures.

Ashley and Ye (2012) applied the Granger causality to identify causal relationship between the median inflation and price dispersion. They observed causal relationship between the median inflation and price dispersion.

Other successful applications of Granger causality were for root cause diagnosis of plant-wide oscillations (Yuan and Joe Qin 2012), between wholesale price and consumer price indices in India (Tiwari 2012) and between tourism and economic growth (Wang et al. 2012).

Akkemik et al. (2012) applied causal analysis to relate energy consumption to income in Chinese provinces. The results obtained indicated that the Chinese government should include a regional viewpoint when articulating and applying energy policies. Wesseh Jr. and Zoumara (2012) studied the causal relationship between energy consumption and economic growth in Liberia whereas Patilea and Raïssi (2012) applied adaptive estimation of vector autoregressive models with time-varying variance to test linear causality in mean.

Suppose we have variables observations $y_k, y_{k-1}, \ldots, y_1$ and $x_{k-1}, x_{k-2}, \ldots, x_i$ then there is a causal relationship between the variables if (Granger 1969):

$$y_k = f\left(\alpha_{k-1} y_{k-1}, \ldots, \alpha_1 y_{1,k-1}, \beta_{k-1} x_{k-1}, \ldots, \beta_1 x_{1,k-1}\right) \qquad (8.4)$$

here α and β are hyper-parameters and are non-zero. In fact from the principles of causality the following factors are valid (Granger 1969):

Observation 1 For there to be causal relationship from variable x to variable y then there has to be correlation relationship between variable y_k and a lagged variable x_q where time q happens before time k.

Observation 2 It is far easier to rule out causality than to observe it.

Observation 3 It is more difficult to know all the variables that cause another variable.

Now that we have described the principles of causality the next step is to create a model that will describe Eq. 8.1 and use it to study the causal relationship between variables and we achieve this goal by implementing the automatic relevance determination method which was described in detail in Chap. 3.

8.6 Automatic Relevance Determination (ARD) for Causality

This chapter proposes automatic relevance determination to study the causality between variables. Automatic relevance determination is a procedure where in a predictive model the causal variable is ranked in order of its effect on the output variable. In this case if there is no causal relationship between variable a and variable b then variable a will be assigned an irrelevancy factor. The application of the ARD within the context of causality studies has not been performed extensively and this chapter proposes the combination of the ARD and Granger causality framework to study the causality between economic variables.

Browne et al. (2008) applied automatic relevance determination for identifying thalamic regions implicated in schizophrenia. The ARD method was created using

Fig. 8.6 The illustration of the automatic relevance determination with the hyperparameters of the prior associated to each input

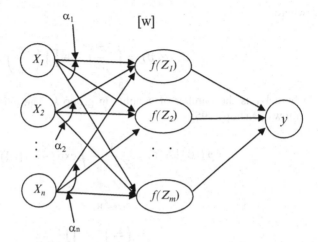

neural networks and the results demonstrated that the thalamic sub-regions were important in schizophrenia. Smyrnakis and Evans (2007) applied the ARD method to classify ischemic events. The ARD method was created using the Bayesian framework and the multilayer perceptron neural network. The ARD technique was applied to classify which of the extracted input features to the multilayer perceptron neural network was the most important with respect to the models performance.

Van Calster et al. (2006) applied the ARD method to classify ovarian tumors. They used the Bayesian multi-layer perceptron networks to choose the most relevant variables. Wang and Lu (2006) applied the ARD technique to approximate urban ozone level and identify the degree of influence by various factors that drive ozone levels. Horner (2005) applied the ARD technique to identify the stage of ovarian cancer in the mass-spectrum of serum proteins whereas Li et al. (2002) applied the ARD method to classify gene expression data and identify relevant features.

The schematic illustration of the ARD is shown in Fig. 8.6.

As described in Chap. 3, an automatic relevance determination technique is created by associating the hyper-parameters of the prior with each input variable and, therefore, the prior can be generalized as follows (MacKay 1991, 1992):

$$E_W = \frac{1}{2} \sum_k \alpha_k \{w\}^T [I_k] \{w\} \tag{8.5}$$

where superscript T is the transpose, k is the weight group and $[I]$ is the identity matrix. As explained in detail in Chap. 3 using the generalized prior in Eq. 8.5, the posterior probability in equation then is (Bishop 1995):

$$P(\{w\}|[D], H_i) = \frac{1}{Z_s} \exp \left(\beta \sum_n \{t_n - y(\{x\}_n)\}^2 - \frac{1}{2} \sum_k \alpha_k \{w\}^T [I_k] \{w\} \right) \tag{8.6}$$

where

$$Z_E(\alpha, \beta) = \left(\frac{2\pi}{\beta}\right)^{\frac{N}{2}} + \prod_k \left(\frac{2\pi}{\alpha_k}\right)^{\frac{W_k}{2}} \tag{8.7}$$

where W_k is the number of weights in group k. The evidence can be written as follows (Bishop 1995):

$$
\begin{aligned}
p\left([D] \mid \alpha, \beta\right) &= \frac{1}{Z_D Z_W} \int \exp\left(-E\left(\{w\}\right)\right) d\{w\} \\
&= \frac{Z_E}{Z_D Z_W} \\
&= \frac{\left(\frac{2\pi}{\beta}\right)^{\frac{N}{2}} + \prod_k \left(\frac{2\pi}{\alpha_k}\right)^{W_k/2}}{\left(\frac{2\pi}{\beta}\right)^{\frac{N}{2}} \prod_k \left(\frac{2\pi}{\alpha_k}\right)^{W_k/2}}
\end{aligned}
\tag{8.8}
$$

Maximization of the log evidence gives the following estimations for the hyper-parameters (Bishop 1995):

$$\beta^{MP} = \frac{N - \gamma}{2 E_D\left(\{w\}^{MP}\right)} \tag{8.9}$$

$$\alpha_k^{MP} = \frac{\gamma_k}{2 E_{W_k}\left(\{w\}^{MP}\right)} \tag{8.10}$$

where $\gamma = \sum_k \gamma_k$, $2 E_{W_k} = \{w\}^T [I_k] \{w\}$ and

$$\gamma_k = \sum_j \left(\frac{\pi_j - \alpha_k}{\eta_j} \left([V]^T [I_k] [V]\right)_{jj}\right) \tag{8.11}$$

and $\{w\}^{MP}$ is the weight vector at the maximum point and this is identified in this chapter using the scaled conjugate gradient method, η_j are the eigenvalues of $[A]$, and $[V]$ are the eigenvalues such that $[V]^T [V] = [I]$.

As described in Chap. 3, the relevance of each input variable, the α_k^{MP} and β^{MP} is estimated by choosing the initial values for the hyper-parameters randomly, training the network using scaled conjugate gradient method to identify $\{w\}^{MP}$ and use Eqs. 8.9 and 8.10 to estimate the hyper parameters and repeating the procedure until convergence without further initialization of the hyper parameters (MacKay 1991). When the α is low then the relevance of a variable on predicting the output variable is high.

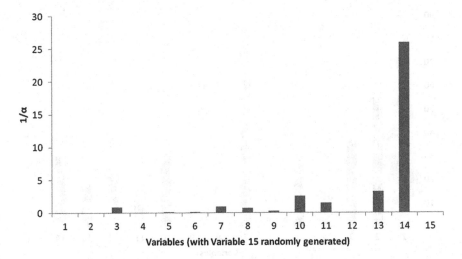

Fig. 8.7 The illustration of the automatic relevance determination with the hyper-parameters of the prior associated to each input

In order to test the method we apply it to the credit rating problem that was discussed in Sect. 8.4. In addition to the 14 input variables we introduce an additional variable which is randomly generated using a uniform distribution. This is to test whether the method will be able to link this variable to the credit scoring. The results of the relevance ($1/\alpha$) are shown in Fig. 8.7. This clearly indicates that there is causal relationship between variable 15 and the credit score.

A further ARD based MLP with the 14 variables as inputs, 8 hidden units and 1 output was constructed and the results of the relevance of each variable on predicting the credit scores is given in Fig. 8.8. This ARD model with this relevance was used to predict the credit scores and the accuracy of 77 % was observed and the resulting receiver operating characteristics (ROC) curve is shown in Fig. 8.9.

The ROC curve is a graphical representation of the sensitivity of the classifier, also known as the true positive rate, against the sensitivity also known as the false positive rate. It has been applied widely for assessing the performance of the classifiers. The ROC has been applied to evaluate the accuracy of classifiers by many researchers and some of the applications include in imaging (Lorimer et al. 2012), for correlated diagnostic marker data (Tang et al. 2012), in network intrusion detection (Meng 2012), field-assessed aerobic fitness related to body size and cardio-metabolic risk in school children (Boddy et al. 2012) and for credit risk rating (Kürüm et al. 2012). The ROC in Fig. 8.9 shows a good classification results. The corresponding confusion matrix is also shown in Table 8.3.

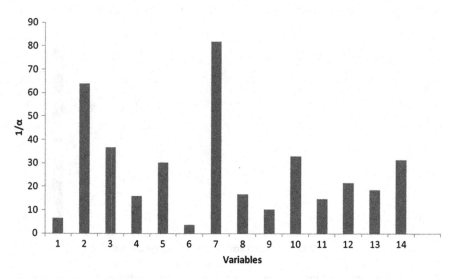

Fig. 8.8 The illustration of the automatic relevance determination with the hyperparameters of the prior associated to each input

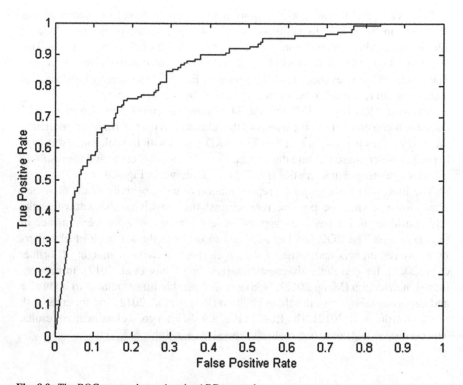

Fig. 8.9 The ROC curve when using the ARD network

Table 8.3 Confusion matrix when the ARD based machine was used

		Actual	
		Positive	Negative
Predicted	Positive	115	34
	Negative	44	147

8.7 Conclusions

This chapter explored the issue of treating a predictive system as a missing data problem i.e. correlation machine and compared it to treating as a cause and effect exercise i.e. causal machine. An auto-associative neural network was combined with genetic algorithm to build the correlation machine. Furthermore, the automatic relevance determination (ARD) approach was also applied to identify causal relationships between variables. These techniques were then applied to modeling the CPI and for credit scoring. The ARD technique was found to be able to asses the causal relationships between the variables while the causal machine was found to perform better than the correlation machine for modeling credit scoring data while the correlation machine was found to perform better than the causal machine for modelling the CPI.

References

Abdella M, Marwala T (2005a) The use of genetic algorithms and neural networks to approximate missing data in database. In: Proceedings of the IEEE 3rd international conference on computational cybernetics, Mauritius, 2005, pp 207–212

Abdella M, Marwala T (2005b) Treatment of missing data using neural networks. In: Proceedings of the IEEE international joint conference on neural networks, Montreal, 2005, pp 598–603

Abdoullaev A (2000) The ultimate of reality: reversible causality. In: Proceedings of the 20th world congress of philosophy, Philosophy Documentation Centre, Boston, internet site, paideia project on-line. http://www.bu.edu/wcp/MainMeta.htm. Last accessed 3 Jan 2013

Achkar R, Owayjan M, Mrad C (2011) Landmine detection and classification using MLP. In: Proceedings of the 3rd international conference on computational intelligence, modeling and simulation, Langkawi, 2011, pp 1–6

Akkemik KA, Göksal K, Li J (2012) Energy consumption and income in Chinese provinces: heterogeneous panel causality analysis. Appl Energy 99:445–454

Akkoç S (2012) An empirical comparison of conventional techniques, neural networks and the three stage hybrid adaptive neuro fuzzy inference system (ANFIS) model for credit scoring analysis: the case of turkish credit card data. Eur J Oper Res 222:168–178

Ashley R, Ye H (2012) On the Granger causality between median inflation and price dispersion. Appl Econ 44:4221–4238

Aydilek IB, Arslan A (2012) A novel hybrid approach to estimating missing values in databases using K-nearest neighbors and neural networks. Int J Innov Comput Inf Control 8:4705–4717

Bishop CM (1995) Neural networks for pattern recognition. Oxford University Press, Oxford

Blanco A, Pino-Mejías R, Lara J, Rayo S (2013) Credit scoring models for the microfinance industry using neural networks: evidence from Peru. Expert Syst Appl 40:356–364

Boddy LM, Thomas NE, Fairclough SJ, Tolfrey K, Brophy S, Rees A, Knox G, Baker JS, Stratton G (2012) ROC generated thresholds for field-assessed aerobic fitness related to body size and cardiometabolic risk in schoolchildren. doi:10.1371/journal.pone.0045755, http://pubget.com/paper/23029224. Last accessed 7 March 2013

Browne A, Jakary A, Vinogradov S, Fu Y, Deicken RF (2008) Automatic relevance determination for identifying thalamic regions implicated in schizophrenia. IEEE Trans Neural Netw 19:1101–1107

Cette G, de Jong M (2013) Breakeven inflation rates and their puzzling correlation relationships. Appl Econ 45:2579–2585

Chen M-H (2010) Pattern recognition of business failure by autoassociative neural networks in considering the missing values. In: Proceedings of the international computer symposium, Taipei, Taiwan, pp 711–715

Chen GG, Åstebro T (2012) Bound and collapse Bayesian reject inference for credit scoring. J Oper Res Soc 63:1374–1387

Chen C-M, Yeh C-Y (2012) The causality examination between demand uncertainty and hotel failure: a case study of international tourist hotels in Taiwan. Int J Hosp Manag 31:1045–1049

Correa AB, Gonzalez AM (2011) Evolutionary algorithms for selecting the architecture of a MLP neural network: a credit scoring case. In: Proceedings of the IEEE international conference on data mining, Vancouver, 2011, pp 725–732

Davis S (2012) Choosing the right baskets for your eggs: deriving actionable customer segments using supervised genetic algorithms. Int J Mark Res 54:689–698

Daya B, Akoum AH, Bahlak S (2012) Geometrical features for multiclass vehicle type recognition using MLP network. J Theory Appl Inf Technol 43:285–294

Dragomir OE, Dragomir F, Brezeanu I, Minc E (2011) MLP neural network as load forecasting tool on short-term horizon. In: Proceedings of the 19th Mediterranean conference on control and automation, Corfu, Greece, pp 1265–1270

Dubey HC, Nandita, Tiwari AK (2012) Blind modulation classification based on MLP and PNN. In: Proceedings of the students conference on engineering and systems, Utter Paredesh, India, art. no. 6199042

Ebrahimpour R, Nikoo H, Masoudnia S, Yousefi MR, Ghaemi MS (2011) Mixture of MLP-experts for trend forecasting of time series: a case study of the Tehran stock exchange. Int J Forecast 27:804–816

García V, Marqués AI, Sánchez JS (2012) Non-parametric statistical analysis of machine learning methods for credit scoring. Adv Intell Syst Comput 171:263–272

Granger CWJ (1969) Investigating causal relations by econometric models and cross-spectral methods. Econometrica 37:424–438

Green C (2003) The lost cause: causation and the mind-body problem. Oxford Forum, Oxford

Hacker RS, Hatemi JA (2006) Tests for causality between integrated variables using asymptotic and bootstrap distributions: theory and application. Appl Econ 38:1489–1500

Horner JK (2005) An automatic relevance determination method for identifying the signature of stage I ovarian cancer in the mass-spectrum of serum proteins. In: Proceedings of the 2005 international conference on artificial intelligence, Las Vegas, Nevada, pp 620–626

Kramer MA (1992) Autoassociative neural networks. Comput Chem Eng 16:313–328

Kürüm E, Yildirak K, Weber G-W (2012) A classification problem of credit risk rating investigated and solved by optimisation of the ROC curve. Cent Eur J Oper Res 20:529–557

Li Y, Campbell C, Tipping M (2002) Bayesian automatic relevance determination algorithms for classifying gene expression data. Bioinformatics 18:1332–1339

Lin Y-J (2012) Comparison of CART- and MLP-based power system transient stability preventive control. Int J Electr Power Energy Syst 45:129–136

Liu L, Wan J (2012) The relationships between Shanghai stock market and CNY/USD exchange rate: new evidence based on cross-correlation analysis, structural cointegration and nonlinear causality test. Physica A Stat Mech Appl 391:6051–6059

Liu J, Gao H, Liu H (2012) Study of optimum solution about aviation ammunition transport problem based on genetic algorithm. In: Proceedings of the IEEE international conference on service operations and logistics, and informatics, Suzhou, China, pp 308–311

Lorimer L, Gemmell HG, Sharp PF, McKiddie FI, Staff RT (2012) Improvement in DMSA imaging using adaptive noise reduction: an ROC analysis. Nucl Med Commun 33:1212–1216

Lu J, Humphreys P, McIvor R, Maguire L, Wiengarten F (2012) Applying genetic algorithms to dampen the impact of price fluctuations in a supply chain. Int J Prod Res 50:5396–5414

MacKay DJC (1991) Bayesian methods for adaptive models. Ph.D. thesis, California Institute of Technology, Pasadena

MacKay DJC (1992) A practical Bayesian framework for back propagation networks. Neural Comput 4:448–472

Marwala T (2009) Computational intelligence for missing data imputation, estimation and management: knowledge optimization techniques. IGI Global Publications, New York

Marwala T (2012) Condition monitoring using computational intelligence methods. Springer, London

Marwala T, Chakraverty S (2006) Fault classification in structures with incomplete measured data using autoassociative neural networks and genetic algorithm. Curr Sci 90:542–548

Meng Y (2012) Measuring intelligent false alarm reduction using an ROC curve-based approach in network intrusion detection. In: Proceedings of the IEEE international conference on computational intelligence for measurement systems and applications, Tianjin, China, pp 108–113

Mistry J, Nelwamondo FV, Marwala T (2009) Investigating demographic influences for HIV classification using Bayesian autoassociative neural networks. Lect Note Comput Sci 5507:752–759

Miyazaki T, Hamori S (2013) Testing for causality between the gold return and stock market performance: evidence for 'gold investment in case of emergency'. Appl Finance Econ 23:27–40

Mutlu Ö, Polat O, Supciller AA (2013) An iterative genetic algorithm for the assembly line worker assignment and balancing problem of type-II. Comput Oper Res 40:418–426

Narayanan S, Marks RJ II, Vian JL, Choi JJ, El-Sharkawi MA, Thompson BB (2002) Set constraint discovery: missing sensor data restoration using auto-associative regression machines. Proc Int Jt Conf Neural Netw 3:2872–2877

Patilea V, Raïssi H (2012) Adaptive estimation of vector autoregressive models with time-varying variance: application to testing linear causality in mean. J Stat Plan Inference 142:2871–2890

Pearl J (2000) Causality: models of reasoning and inference. Cambridge University Press, Cambridge

Pereira LFA, Pinheiro HNB, Silva JIS, Silva AG, Pina TML, Cavalcanti GDC, Ren TI, De Oliveira JPN (2012) A fingerprint spoof detection based on MLP and SVM. In: Proceedings of the international IEEE joint conference on neural networks. Brisbane, Australia, 2012, art. no. 6252582

Peyghami MR, Khanduzi R (2012) Predictability and forecasting automotive price based on a hybrid train algorithm of MLP neural network. Neural Comput Appl 21:125–132

Quinlan JR (1987) Simplifying decision trees. Int J Man Mach Stud 27:221–234

Quinlan JR (1992) C4.5: programs for machine learning. Morgan Kaufmann, San Mateo

Ramírez Palencia AE, Mejía Delgadillo GE (2012) A computer application for a bus body assembly line using genetic algorithms. Int J Prod Econ 140:431–438

Rezaeian-Zadeh M, Tabari H (2012) MLP-based drought forecasting in different climatic regions. Theory Appl Climatol 109:407–414

Salazar DSP, Adeodato PJL, Arnaud AL (2012) Data transformations and seasonality adjustments improve forecasts of MLP ensembles. In: Proceedings of the IEEE conference on evolving and adaptive intelligent systems, Madrid, Spain, pp 139–144

Sermpinis G, Dunis C, Laws J, Stasinakis C (2012) Forecasting and trading the EUR/USD exchange rate with stochastic neural network combination and time-varying leverage. Decis Support Syst 54:316–329

Sioud A, Gravel M, Gagné C (2012) A hybrid genetic algorithm for the single machine scheduling problem with sequence-dependent setup times. Comput Oper Res 39:2415–2424

Smyrnakis MG, Evans DJ (2007) Classifying ischemic events using a Bayesian inference multilayer perceptron and input variable evaluation using automatic relevance determination. Comput Cardiol 34:305–308

Tang B, Qiu S (2012) A new credit scoring method based on improved fuzzy support vector machine. In: Proceedings of the IEEE international conference on computer science and automation engineering, pp 73–75

Tang LL, Liu A, Chen Z, Schisterman EF, Zhang B, Miao Z (2012) Nonparametric ROC summary statistics for correlated diagnostic marker data. Stat Med. doi:10.1002/sim.5654

Tiwari AK (2012) Causality between: an empirical investigation in the frequency domain. Indian Growth Dev Rev 5:151–172

Turchenko V, Beraldi P, De Simone F, Grandinetti L (2011) Short-term stock price prediction using MLP in moving simulation mode. In: Proceedings of the 6th IEEE international conference on intelligent data acquisition and advanced computing systems: technology and applications, Prague, 2011, pp 666–671

Van Calster B, Timmerman D, Nabney IT, Valentin L, Van Holsbeke C, Van Huffel S (2006) Classifying ovarian tumors using Bayesian multi-layer perceptrons and automatic relevance determination: a multi-center study. Proc Annu Int Conf IEEE Eng Med Biol Soc 1:5342–5345

Vedala R, Kumar BR (2012) An application of Naive Bayes classification for credit scoring in e-lending platform. In: Proceedings of the international conference on data science and engineering, Kerala, India, pp 81–84

Vidal T, Crainic TG, Gendreau M, Prins C (2013) A hybrid genetic algorithm with adaptive diversity management for a large class of vehicle routing problems with time-windows. Comput Oper Res 40:475–489

Vieira A, Ribeiro B, Chen N (2012) Credit scoring for SME using a manifold supervised learning algorithm. Lect Note Comput Sci 7435:763–770

Vladislavleva E, Friedrich T, Neumann F, Wagner M (2012) Predicting the energy output of wind farms based on weather data: important variables and their correlation. Renew Energy 50:236–243

Wang D, Lu W-Z (2006) Interval estimation of urban ozone level and selection of influential factors by employing automatic relevance determination model. Chemosphere 62:1600–1611

Wang L, Zhang H, Li W (2012) Analysis of causality between tourism and economic growth based on computational econometrics. J Comput 7:2152–2159

Wesseh PK Jr, Zoumara B (2012) Causal independence between energy consumption and economic growth in Liberia: evidence from a non-parametric bootstrapped causality test. Energy Policy 50:518–527

Yuan T, Joe Qin S (2012) Root cause diagnosis of plant-wide oscillations using granger causality. IFAC Proc Vol (IFAC-Pap Online) 8(Part 1):160–165

Zieba M, Światek J (2012) Ensemble classifier for solving credit scoring problems. IFIP Adv Inf Commun Technol 372:59–66

Chapter 9
Evolutionary Approaches to Computational Economics: Application to Portfolio Optimization

Abstract This chapter examines the use of genetic algorithms (GA) to perform the task of continuously rebalancing a portfolio, targeting specific risk and return characteristics. The portfolio is comprised of a number of arbitrarily performing trading strategies, plus a risk-free strategy in order to rebalance in a way similar to the traditional Capital Asset Pricing Model (CAPM) method of rebalancing portfolios. A format is presented for the design of a fitness function appropriate to the task, which is evaluated by examining the final results. The results of targeting both risk and return were investigated and compared, as well as optimizing the non-targeted variable to create efficient portfolios. The findings showed that GA is, indeed, a viable tool for optimizing a targeted portfolio using the presented fitness function.

9.1 Introduction

This chapter, applies genetic algorithm (GA) to continuously re-balance a portfolio by targeting specific risk and return characteristics. This chapter is based on the work of Hurwitz and Marwala (2012). The portfolio is comprised of a number of arbitrarily performing trading strategies, plus a risk-free strategy to rebalance in a similar method to the traditional Capital Asset Pricing Model (CAPM) method of rebalancing portfolios. A format is presented for designing a fitness function appropriate to the task, and evaluated by examining the final results. The results of targeting both the risk and return were investigated and compared. In addition, the non-targeted variable was optimized to create efficient portfolios. The viability of seeking a targeted portfolio using a GA and the proposed fitness function was evaluated.

In the course of managing funds, it is often required of the fund manager to manage funds in such a manner as to obtain specific measurable bounds of the portfolio's performance. In particular, the two most commonly used measures are that of *return* and *risk*. The need for such targeted funds are many and varied, but

T. Marwala, *Economic Modeling Using Artificial Intelligence Methods*, Advanced Information and Knowledge Processing, DOI 10.1007/978-1-4471-5010-7_9, © Springer-Verlag London 2013

can all be expressed by having either a targeted, normally minimum, return, with a minimized risk value, or targetting a maximum risk value with a maximized return value (Fischer and Jordan 1991). The theoretical model originally proposed by Markowitz (1952) is used to model all available investments, including composite investment strategies, in terms of their historical risks and returns. These are combined to form an efficient portfolio, which forms the basis of the financial engineering used to gain a targeted return. This chapter explores the use of a GA to perform that balancing, targeting both risk, as measured as a statistical variance, and return. The use of a GA to optimize non-linear multi-variable problems is well explored. This problen can be framed as an optimization problem and a GA can be used to optimize risks and rewards.

Optimization in this chapter is intended to solve the traditional capital asset pricing model problem of achieving a targeted fund. This is done by including a risk-free asset into the suite of strategies for the GA to optimize with. The methods examined in this chapter are for creating a targeted risk or return portfolio in an uncertain environment, where the characteristics of each strategy change over time, thus requiring constant rebalancing.

A successful implementation of GA to portfolio optimization was conducted by Chang et al. (2009) for different risk measures. The GA was compared to a mean–variance model in a cardinality constrained efficient frontier. Chang et al. (2009) collected three different risk measures based on a mean–variance model by Markowitz, together with semi-variance, mean absolute deviation, and variance with skewness. They demonstrated that these portfolio optimization problems can be solved through the use of a GA if mean–variance, semi-variance, mean absolute deviation and variance with skewness are used as the measures of risk. The robustness of the GA technique was verified by three data sets collected from financial markets. The results demonstrated that the investors should include only one third of their total assets in the portfolio which then outperforms those that contained more assets.

Another application of GA was conducted by Chen et al. (2011) who applied a particular variant with guided mutation called Genetic Relation Algorithm (GRA) for large-scale portfolio optimization. GRA is one of the evolutionary approaches that have a graph structure. Chen et al. (2011) introduced a new operator, called *guided mutation*. To select the most efficient portfolio, GRA considers the correlation coefficient between stock brands as strength, which shows the relation between nodes in each individual of the GRA. Guided mutation produces offspring according to the average value of correlation coefficients of each individual, to enhance the exploitation ability of evolution of the GRA. When applied, the results showed that the GRA approach was successful in portfolio optimization. This study demonstrates that sometimes GA ought to be modified in order to obtain good results.

A successful application of GA for portfolio optimization was conducted by Oh et al. (2005) who applied it to index fund management. *Index fund management* is one of the popular strategies in portfolio management and its objective is to match the performance of the benchmark index such as the FTSE 100 in London.

This strategy is applied by fund managers particularly when they are not sure about outperforming the market and want to adjust themselves to average performance. The application of GA in this problem gave good results.

A successful application of GA was conducted by Lin and Liu (2008) applied it to portfolio selection problems with minimum transaction lost. Usually, portfolio selection problems are solved with quadratic or linear programming models. However, the solutions obtained by these approaches are in real numbers and difficult to apply because each asset usually has its minimum transaction lot. Approaches that consider minimum transaction lots were developed based on some linear portfolio optimization models. No study has ever investigated the minimum transaction lot problem in portfolio optimization based on Markowitz' model, which is probably the most well-known and widely used. Based on Markowitz' model, Lin and Liu (2008) presented three possible models for portfolio selection problems with minimum transaction lots, and devised corresponding GA models to obtain the solutions. The results of Lin and Liu (2008) showed that the portfolios obtained using the proposed algorithms were very close to the efficient frontier, indicating that the presented method can obtain near optimal results and find practically feasible solutions to the portfolio selection problem in an acceptably short time. They recommended their model based on a fuzzy multi-objective decision-making framework because of its adaptability and simplicity.

Other successful applications of genetic algorithms to portfolio optimization include Anagnostopoulos and Mamanis (2011) who applied a GA in the mean–variance cardinality constrained portfolio optimization problem as well as Chen et al. (2010) who applied time-adapting genetic-network programming for portfolio optimization.

9.2 Background

This section covers the background theory of portfolio optimization and that of optimization through the use of a genetic algorithm (GA). In the case of portfolio optimization, the basic theory assumes a stationary system, an assumption that if we discard, and it is desirable to discard, needs to be accounted for (Hurwitz 2012; Hurwitz and Marwala 2011, 2012).

9.2.1 Modern Portfolio Theory

To characterize a given investment portfolio, it is measured according to both its *return* and its *risk* characteristics (Markowitz 1952). Over a given period of time, the *return* of a portfolio (R) is characterized by Eq. 9.1 (Markowitz 1952), which determines the percentage gain (or loss) made by the portfolio over a given period.

$$R_P = I_{RF} + (R_M - I_{RF})\sigma_P/\sigma_M \qquad (9.1)$$

Fig. 9.1 Return-risk and the
efficient horizon

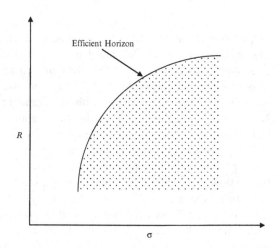

Here R_P is the Expected Return of Portfolio, R_M is the Return on the Market Portfolio, I_{RF} is the Risk-Free rate of Interest, σ_M is the Standard Deviation of the Market Portfolio, and σ_P is the Standard Deviation of the Portfolio.

The *risk* of a given portfolio over a set period of time is measured by obtaining its statistical variance (Markowitz 1952). These two variables then become the characteristic measurements of a portfolio's performance (Markowitz 1952). To compare portfolios, the portfolios are plotted together on a Cartesian plane, plotting *return*, R versus *risk*, σ, as shown in Fig. 9.1.

Note the characteristic line, marking what is referred to as the *Efficient Horizon*. This frontier line marks the minimal values of risk for a given reward obtainable with the given portfolios, or conversely, the maximum return obtainable for a given value of risk. If a portfolio lies on this line, it is said to be *efficient* (Markowitz 1952).

9.2.2 CAPM Modeling

CAPM (Capital Asset Pricing Model) modeling is a process designed to obtain a given value of risk from a composite portfolio, often significantly lower in value than the lowest efficient portfolio in the universe of portfolios (Jiang 2011). To achieve this, an *optimal portfolio* is first found along the efficient horizon (Jiang 2011). This portfolio is then combined in a linear fashion with a *zero-risk* portfolio in a proportional manner to create a new portfolio, which meets the required specifications (Jiang 2011). This process is illustrated in Fig. 9.2.

A *zero-risk* asset is an asset so safe that it has a variance of zero (Jiang 2011). Such commonly used assets are treasury bonds, and often other government-backed securities that are assumed never to default. This obviously does not account for political risk, or the possibility of a government going bankrupt, something that in all likelihood needs a deeper examination, considering the recent global credit

Fig. 9.2 CAPM portfolio
targeting

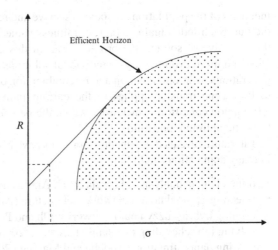

crisis, and the on-going effects still being felt. In this way, the system is analogous
to a feed-forward control system, that aims to control the targeted variable (return
or risk) while, at the same time, optimizing the secondary variable.

9.2.3   Genetic Algorithms

A genetic algorithm (GA) uses the principles of genetic evolution to optimize multi-
variable, non-linear problems (Holland 1975). The heart of the GA is in the concept
of the *fitness function* (Holland 1975), which evaluates any given solution, and
provides a score, based on that solution. The algorithm creates a population of
possible solutions, and then pairs likely breeders from the population based on
their fitness. The breeding pairs then combine their various values (referred to in
the literature as *chromosomes*) through various methods (Holland 1975). To avoid
becoming stuck in local minima, the GA employs one of a number of mutation
operations (again, based on biological genetic mutation), giving the population
enough diversity to find solutions outside of the local minima (Holland 1975). Being
an evolutionary optimization technique, there is no guarantee of finding a global
optimum, although in most cases this limitation is purely academic (Hart 1994).

Unlike many optimization algorithms, a genetic algorithm has a higher probability of converging to a global optimal solution than a gradient-based method.
A genetic algorithm is a population-based, probabilistic technique that operates
to find a solution to a problem from a population of possible solutions (Kubalík
and Lazanský 1999). It is used to find approximate solutions to difficult problems
through the application of the principles of evolutionary biology to computer science
(Michalewicz 1996; Mitchell 1996; Forrest 1996; Vose 1999; Marwala 2003, 2010;
Tettey and Marwala 2006). It is analogous to Darwin's theory of evolution where

members of the population compete to survive and reproduce while the weaker ones die out. Each individual is assigned a fitness value according to how well it meets the objective of solving the problem and, in this chapter, portfolio optimization. New and more evolutionary-fit individual solutions are produced during a cycle of generations, wherein selection and re-combination operations take place, analogous to how gene transfer applies to the current individuals. This continues until a termination condition is met, after which the best individual thus far is considered to be the optimum portfolio.

Literature reveals that genetic algorithms have been proven to be very successful in many applications including:

- finite element analysis (Marwala 2002; Akula and Ganguli 2003),
- selecting optimal neural network architecture (Arifovic and Gençay 2001),
- training hybrid fuzzy neural networks (Oh and Pedrycz 2006),
- solving job scheduling problems (Park et al. 2003),
- missing data estimation (Abdella and Marwala 2006), and
- portfolio optimization (Chen et al. 2009a, b; Soleimani et al. 2009; Bermúdez et al. 2012; Branke et al. 2009).

Furthermore, the genetic algorithm method has been proven to be successful in complex optimization problems such as wire routing, scheduling, adaptive control, game playing, cognitive modeling, transportation problems, traveling salesman problems, optimal control problems, and database query optimization (Pendharkar and Rodger 1999; Marwala et al. 2001; Marwala and Chakraverty 2006; Marwala 2007; Crossingham and Marwala 2007; Hulley and Marwala 2007).

In this chapter, the genetic algorithm views learning as a competition among a population of evolving candidate problem solutions. A fitness function evaluates each solution to decide whether it will contribute to the next generation of solutions. Through operations analogous to gene transfer in sexual reproduction, the algorithm creates a new population of candidate solutions (Goldberg 1989). The three most important aspects of using genetic algorithms are:

1. the definition of the objective function;
2. implementation of the genetic representation; and
3. implementation of the genetic operators (initialization, crossover, mutation, selection, and termination).

9.2.3.1 Initialization

In the beginning, a large number of possible individual solutions are randomly generated to form an initial population. This initial population is sampled so that it covers a good representation of the updating solution space. For example, if there are two variables to be updated, the size of the population must be greater than when there is only one variable to be updated.

9.2.3.2 Crossover

The crossover operator mixes genetic information in the population by cutting pairs of chromosomes at random points along their length and exchanging the cut sections over. This has a potential for joining successful operators together. Crossover, in this chapter, is an algorithmic operator used to alter the programming of a potential solution to the optimization problem from one generation to the other (Gwiazda 2006).

Crossover occurs with a certain probability. In many natural systems, the probability of crossover occurring is higher than the probability of mutation occurring. One example is a simple crossover technique (Goldberg 1989).

For *simple crossover*, one crossover point is selected, a binary string from the beginning of a chromosome to the cross-over point is copied from one parent, and the rest is copied from the second parent. For example, if two chromosomes in binary space are $a = 11001011$ and $b = 11011111$ and they undergo a *one-point crossover* at the midpoint, then the resulting offspring may be $c = 11001111$. For an *arithmetic crossover*, a mathematical operation is additionally performed to make the offspring. For example, an AND operator can be performed on $a = 11001011$ and $b = 11011111$ to form an offspring $d = 11001011$. For this chapter, arithmetic crossover was used.

9.2.3.3 Mutation

The mutation operator picks a binary digit of the chromosomes at random and inverts it. This has a potential of introducing new information to the population, and, thereby, prevents the genetic algorithm simulation from being stuck in a local optimum. Mutation occurs with a certain probability. In many natural systems, the probability of mutation is low (normally less than 1 %). In this chapter, binary mutation is used (Goldberg 1989). When binary mutation is used, a number written in binary form is chosen and one bit value is inverted. For example: the chromosome $a = 11001011$ may become the chromosome $b = 11000011$.

9.2.3.4 Selection

For every generation, a selection of the proportion of the existing population is chosen to breed a new population. This selection is conducted using the fitness-based process, where solutions that are fitter are given a higher probability of being selected. Some selection methods rank the fitness of each solution and choose the best solutions while other procedures rank a randomly chosen sample of the population for computational efficiency.

Many selection functions tend to be stochastic in nature and thus are designed in such a way that a selection process is conducted on a small proportion of less fit solutions. This ensures that diversity of the population of possible solutions is maintained at a high level and, therefore, avoids converging on poor and incorrect solutions. There are many selection methods and these include roulette-wheel selection (Mohamed et al. 2008).

Roulette-wheel selection is a genetic operator used for selecting potentially useful solutions in a genetic optimization process. In this method, each possible procedure is assigned a fitness function which is used to map the probability of selection with each individual solution. Suppose the fitness f_i is of individual i in the population, then the probability that this individual is selected is (Goldberg 1989):

$$p_i = \frac{f_i}{\sum\limits_{j=1}^{N} f_j} \tag{9.2}$$

where N is the total population size.

This process ensures that candidate solutions with a higher fitness have a lower probability so that they may eliminate those with a lower fitness. By the same token, solutions with low fitness have a low probability of surviving the selection process. The advantage of this is that, even though a solution may have low fitness, it may still contain some components which may be useful in the future.

The processes described result in the subsequent generation of a population of solutions that is different from the previous generation and that has an average fitness that is higher than the previous generation.

9.2.3.5 Termination

The process described is repeated until a termination condition has been achieved, either because a desired solution that satisfies the fitness function was found or because a specified number of generations has been reached or the solution's fitness has converged (or any combination of these).

The process described above is shown in pseudo code in Algorithm 9.1 (Goldberg 1989). Table 9.1 shows the operations, types, and parameters in the implementation of a genetic algorithm. This table indicates that in a genetic algorithm implementation, there are many choices that ought to be made. For example, for genetic algorithm representation, a choice has to be made between a binary and a floating point representation. For this chapter, a binary representation was used. Given this choice, a bit size must be chosen. For this chapter, a 16-bit binary representation was chosen.

Algorithm 9.1 An algorithm for implementing a genetic algorithm (reprinted with permission from Marwala 2009)

- Select the initial population
- Calculate the fitness of each chromosome in the population:
- Repeat

    - Choose chromosomes with higher fitness to reproduce.
    - Generate a new population using crossover and mutation to produce offspring.
    - Calculate the fitness of each offspring.
    - Replace the low fitness section of the population with offspring .

- Until termination

Table 9.1 Operations, types and parameters for the implementation of a genetic algorithm

Operation	Types	Parameters to select
Genetic representation	Binary, floating point	Bit size
Initialization	Population, random seed	Population size, distribution of the random seed
Crossover	Arithmetic, simple, one-point, two-point, uniform	Probability of crossover
Mutation	Non-uniform, binary	Probability of mutation
Fitness function evaluation	Problem specific	
Selection	Roulette wheel, tournament selection	
Reproduction	Two parents, three parents reproduction	

For the initialization process, a choice has to be made for the population size. Table 9.1 illustrates that in the implementation of genetic algorithms, the difficulty is that there are many choices to be made and there is no direct methodology on how these choices must be made, so these choices tend to be arbitrary.

9.3 Problem Statement

To investigate the problem, a number of simple trading strategies were required. To ensure that enough facets were available to trade between, data was generated with specific underlying patterns (Hurwitz and Marwala 2011), and then was overlaid with an energy-based noise function (Hurwitz and Marwala 2011). These datasets (or pseudo-data) were then traded on using a simple trading strategy that was based upon their underlying facets (Hurwitz and Marwala 2011). These trading strategies were deliberately not fine-tuned, as there needs to be periods of loss and gain, as well as low and high risk in each trading strategy. A final strategy was then created, being a buy-and-hold strategy that invested in a low-return bond, emulating a *zero-risk* portfolio (Hurwitz and Marwala 2012).

The strategies were not perfect exploiters of their data-sets. In many cases they performed admirably, while in others they actually lost money, owing to the individual idiosyncrasies of their trading strategy interacting with the data they were attempting to exploit. A case in point is the first strategy, which performed well on a steep gradient, but performed very poorly when on a flat gradient, being fooled into poor trades by the noise in the data.

9.4   Genetic Algorithm Setup

To use the genetic algorithm (GA) to optimize the portfolio, a number of values must be specified. The first is the form of the individual, which is characterized by four variables, each representing the relative weighting of each strategy. Since the GA has freedom to use any combination of values within the set bounds (in this case the set between 0 and 1), the weights must be rescaled to account for all of the funds held by the portfolio. This rescaling is done as shown in Eq. 9.3 (Hurwitz and Marwala 2012).

$$W_i = \frac{W_i}{\sum_{j=1}^{n} w_j} \tag{9.3}$$

The rescaled values were also enforced on the individuals, aiding in convergence. The next step involved the evaluation of the individual fitness from a given individual. To do this, first the actual return and risk for the given period was calculated for each strategy. Using the given weighting factors, the risk and return for the combined portfolio represented by the individual were calculated. These were then compared to the target value, and the error found was the primary component of the fitness function. To optimize the portfolio, the secondary factor was then used to either maximize the return of the targeted risk portfolio, or to minimize the risk of a targeted return portfolio. Equations 9.4 and 9.5 show these two fitness functions respectively, where σ is the risk, R is the return, and T is the target value (Hurwitz and Marwala 2012).

$$F_\sigma = -k \left(\left| [D][\sigma'] - T \right| \right) + [D][R'] \tag{9.4}$$

$$F_R = -k \left(\left| [D][R'] - T \right| \right) - [D][\sigma'] \tag{9.5}$$

Note that in both cases, the factor k determines the amount by which the primary target takes precedence over optimizing the secondary target. By adjusting k, some slight accuracy can, in theory, be sacrificed to obtain overall better performance. Equations 9.4 and 9.5 can also be modified so that the first term returns no error if the performance exceeds the bound set by the target (so it returns zero if $[D][\sigma']$ is less than T, or if $[D][R']$ is greater than T), creating the following equations (Hurwitz and Marwala 2012).

$$if\ [D]\left[\sigma'\right] > T;\ F_\sigma = -k\left(\left|[D]\left[\sigma'\right] - T\right|\right) - [D]\left[R'\right]$$
$$else\ F_\sigma = [D]\left[R'\right] \tag{9.6}$$

$$if\ [D]\left[R'\right] < T;\ F_R = -k\left(\left|[D]\left[R'\right] - T\right|\right) - [D]\left[\sigma'\right]$$
$$else\ F_R = -[D]\left[R'\right] \tag{9.7}$$

By doing so, the GA is not forced to (potentially) sacrifice performance to meet an exact target specification, but is instead free to perform within acceptable bounds so as to achieve better overall performance.

9.5 Analysis

This section examines the details of the optimization process, covering the technical specifications of the optimizations and gives an analysis of the results.

9.5.1 Technical Specifications

All three strategies used a traditional Moving Average Convergence-Divergence (MACD) indicator for generating their buy and sell indicators (Kilpatrick and Dahlquist 2010). The portfolio trading was limited to trading no more than once a week, with daily values used to determine the results. In accordance with common reporting standards (Dodel 2011), risk and return results were computed on a monthly basis, since it is this target that a targeted risk-return fund would be aiming for (Jiang 2011).

The first experiment was to attempt to create a targeted risk fund, one easily achievable with the given strategies, namely a *zero-risk* fund. This should be very easy to achieve, as the GA only needs to choose to use the risk-free asset exclusively. It is, however, not completely risk-free, as the growth is based on a yearly return, and the daily returns have a slight variance owing to the non-reported days in between, such as weekends and public holidays. The first and most notable issue is that the performance is incredibly poor, owing to its natural preference to choose strategies that are currently sold, and hence have a genuinely *zero-risk* factor and correspondingly zero return. To avoid this trap, it proved necessary to introduce a large penalty for favoring any dis-invested strategy. With this done, the simulations could continue unimpeded. Multiple portfolios were then created, using varying targets and parameter settings and the parameters in question were the factor k and the choice of whether to aim for an exact targeted risk and return or simply set an upper or lower bound on the risk and return targets. Portfolios were generated targeting both risk and return, with their performance measured against their targets, and relatively against each other using the non-targeted, optimized measure. Errors were expected to be present, owing to the system's reliance on historical data to predict future events. The ability to spread the risk between multiple strategies was intended to mitigate this problem somewhat, although it is unreasonable to expect it to be fully compensated for.

9.5.2 Results Analysis

The first set of portfolios looked at targeted risk as the control variable. The targeting of the variable is a hard target, desiring an exact controlled match. This first portfolio was designed to obtain the minimum possible risk that the component strategies allowed. The results shown in Fig. 9.3 indicated that the portfolio achieved the desired risk, or better, for the vast majority of the time period. Spikes of poor performance are inevitable, owing to the lack of any feedback in the system, and the unreliable nature of the component strategies themselves.

When studying the strategy choices made by GA when optimizing the portfolio, it was observed that there was the heavy dependence (as expected) on the *zero-risk* portfolio. The rest of the portfolios were generated in a similar manner, with the various parameters changing. Table 9.2 shows the targeted risk portfolio as well as the parameter settings used in each instance.

Some portfolios have been omitted to preserve a measure of clarity in the figures. Of particular interest is the fact that as the targeted risk is relaxed (i.e., set to a higher value), the returns achieved by the portfolio become greater, matching the expectations of modern portfolio theory. The achievement of a mean variance within the targeted risk bounds indicates the usefulness of the technique to successfully balance (and rebalance) a portfolio. Of further interest is the difference between portfolios with the same targeted risk, but where one is allowed for the risk to be merely an upper bound, while the other targets the exact value. As can be seen, the freedom to not specifically target a given risk value actually allows for a greater return, while the mean risk still remains within the controlled bounds. Another facet worth mentioning is the effect of the factor k, we can see portfolios number 2 and number 5 perform the same function, but while number 2 has k set to 100, number 5 is far less stringent, setting its value to only 10. The resulting portfolio in number 5 has a better return than number 2, but a more wildly fluctuating risk, the variance of the risk is sacrificed in exchange for a better return.

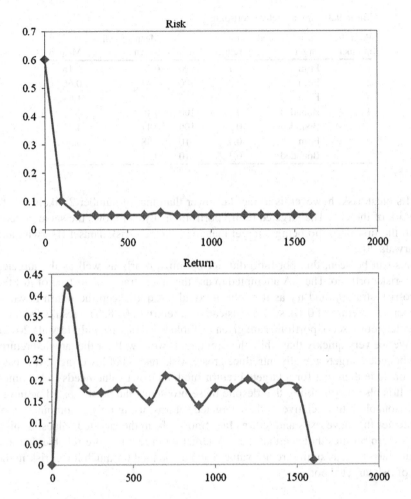

Fig. 9.3 Minimum-risk scenario

Table 9.2 Targeted risk portfolio

Portfolio number	Firm/bounded target	Target σ	K	Mean σ	Mean periodic return
1	Firm	0.15	100	0.022	0.15
2	Firm	0.2	100	0.44	0.27
3	Firm	1	100	0.93	0.47
4	Bounded	1	10	0.55	0.58
5	Bounded	0.2	10	0.073	0.29
6	Firm	1	10	0.92	0.45
7	Bounded	0.2	100	0.069	0.24

Table 9.3 Targeted return portfolios

Portfolio number	Firm/bounded target	Target return	K	Mean periodic return	Mean σ
1	Firm	0.1	100	0.28	0.16
2	Firm	0.3	100	0.43	0.66
3	Firm	0.5	100	0.58	1.04
4	Bounded	0.1	100	0.40	4.2
5	Bounded	0.3	100	0.44	4.3
6	Firm	0.3	10	0.38	0.34
7	Bounded	0.3	10	0.31	3.6

Its mean risk, however, is in fact far lower than that of number 2, likely due to a quirk of the GA. The targeted return portfolio analysis follows the same pattern, with the first portfolio being a target matching the zero-risk dataset (Hurwitz and Marwala 2012).

As can be seen, this portfolio did not optimize nearly as well as the targeted zero-risk portfolio. The GA attempted to use the other strategies, instead of just the zero-risk strategy, and try as it might, it could not quite keep the return down to the targeted return of 0.16, settling instead on a return of 0.28. The results from the targeted return set of portfolios are given in Table 9.3 (Hurwitz and Marwala 2012).

We see very quickly that while the firm targets work well for the targeted return, the bounded targets actually introduce greater risk, instead of lowering it. It is easy to conclude then that for a targeted return fund, an exact value needs to be aimed for. It is also worth noting that despite these problems, the portfolios all achieved a reasonable return relative to their targeted returns, using only a combination of strategies that have risks and returns far-removed from the targeted values. Similar risk-return profiles indicate that the risk-targeting in fact is more reliable than the return targeting, as similar return values can be obtained for much lower risk in the set of risk targeted portfolios.

9.6   Conclusions

As shown, the use of genetic algorithms for the continual rebalancing of portfolios is a viable technique that holds much promise. Given a set of strategies, the GA was able to choose between them to meet a given target of performance, be it risk or reward, and to optimize the secondary target. The inclusion of a risk-free asset into the optimization process obviates the need to first find an optimal, efficient portfolio to then linearly combine with the risk-free asset, instead of allowing the system to move directly on to balancing a targeted portfolio that meets the investor's needs. The relaxing of precise control restraints in favor of bounded control restraints has proven to be a highly promising technique, allowing for improved performance whilst still meeting the target performance requirements. It was found that risk targeting produces a more reliable and efficient portfolio optimization strategy than risk targeting, which may be due to the data-generation method. The inherent unreliability of the various strategies, while problematic, proved not to sabotage the performance of the overall portfolio, although it did, unsurprisingly, heighten the element of risk. Nonetheless, this risk was manageable within acceptable bounds. For more stable results, it is recommended that a penalty be introduced to the fitness function for more radical changes, thereby, favoring strategies that limit their amount of trading (also bringing down trading costs for real-world application). Further work is recommended using combinations of both more advanced trading rules, and even simple buy-and-hold strategies, both on idealized data and on real-world equity data.

References

Abdella M, Marwala T (2006) The use of genetic algorithms and neural networks to approximate missing data in database. Comput Inform 24:1001–1013

Akula VR, Ganguli R (2003) Finite element model updating for helicopter rotor blade using genetic algorithm. AIAA J 41(3):554–556. doi:10.2514/2.1983

Anagnostopoulos KP, Mamanis G (2011) The mean–variance cardinality constrained portfolio optimization problem: an experimental evaluation of five multiobjective evolutionary algorithms. Expert Syst Appl 38:14208–14217

Arifovic J, Gençay R (2001) Using genetic algorithms to select architecture of a feedforward artificial neural network. Physica A Stat Mech Appl 289:574–594

Bermúdez JD, Segura JV, Vercher E (2012) A multi-objective genetic algorithm for cardinality constrained fuzzy portfolio selection. Fuzzy Set Syst 188:16–26

Branke J, Scheckenbach B, Stein M, Deb K, Schmeck H (2009) Portfolio optimization with an envelope-based multi-objective evolutionary algorithm. Eur J Oper Res 199:684–693

Chang TJ, Yang S-C, Chang K-J (2009) Portfolio optimization problems in different risk measures using genetic algorithm. Expert Syst Appl 36:10529–10537

Chen AJ-S, Hou J-L, Wu S-M, Chang-Chien Y-W (2009a) Constructing investment strategy portfolios by combination genetic algorithms. Expert Syst Appl 36:3824–3828

Chen Y, Ohkawa E, Mabu S, Shimada K, Hirasawa K (2009b) A portfolio optimization model using genetic network programming with control nodes. Expert Syst Appl 36:10735–10745

Chen Y, Mabu S, Hirasawa K (2010) A model of portfolio optimization using time adapting genetic network programming. Comput Oper Res 37:1697–1707

Chen Y, Mabu S, Hirasawa K (2011) Genetic relation algorithm with guided mutation for the large-scale portfolio optimization. Expert Syst Appl 38:3353–3363

Crossingham B, Marwala T (2007) Using genetic algorithms to optimise rough set partition sizes for HIV data analysis. Stud Comput Intell 78:245–250

Dodel K (2011) Financial reporting and analysis. CFA Institute report, CFA Institute, Charlottesville

Fischer DE, Jordan RJ (1991) Security analysis and portfolio management. Prentice Hall, Englewood cliffs

Forrest S (1996) Genetic algorithms. ACM Comput Surv 28:77–80

Goldberg DE (1989) Genetic algorithms in search, optimization and machine learning. Addison-Wesley, Reading

Gwiazda TD (2006) Genetic algorithms reference vol 1 cross-over for single-objective numerical optimization problems. Adobe eBook, Lomianki

Hart WE (1994) Adaptive global optimization with local search. Ph.D. thesis, University of California, San Diego

Holland JH (1975) Adaptation in natural and artificial systems. University of Michigan Press, Ann Arbor

Hulley G, Marwala T (2007) Genetic algorithm based incremental learning for optimal weight and classifier selection. Comput Model Life Sci Am Inst Phys Ser 952:258–267

Hurwitz E (2012) Portfolio optimization. Doctoral thesis in preparation, University of Johannesburg, Johannesburg

Hurwitz E, Marwala T (2011) Suitability of using technical indicator-based strategies as potential strategies within intelligent trading systems. In: Proceedings of the IEEE international conference on systems, man, and cybernetics, Anchorage, 2011, pp 80–84

Hurwitz E, Marwala T (2012) Optimising a targeted fund of strategies using genetic algorithms. In: Proceedings of the IEEE international conference on systems, man, and cybernetics, Seoul, 2012, pp 2139–2143

Jiang P (2011) Corporate finance and portfolio management. CFA Institute report, CFA Institute, Charlottesville

Kilpatrick CD II, Dahlquist JR (2010) Technical analysis: the complete resource for financial market technicians. Pearson, London

Kubalík J, Lazanský J (1999) Genetic algorithms and their testing. AIP Conf Proc 465:217–229

Lin C-C, Liu Y-T (2008) Genetic algorithms for portfolio selection problems with minimum transaction lots. Eur J Oper Res 185:393–404

Markowitz H (1952) Portfolio selection. J Finance 7:77–91

Marwala T (2002) Finite element updating using wavelet data and genetic algorithm. AIAA J Aircr 39:709–711

Marwala T (2003) Control of fermentation process using Bayesian neural networks and genetic algorithm. In: Proceedings of the African control conference, Cape Town, 2003, pp 449–454

Marwala T (2007) Bayesian training of neural network using genetic programming. Pattern Recognit Lett 28:1452–1458

Marwala T (2009) Computational intelligence for missing data imputation, estimation and management: knowledge optimization techniques. IGI Global Publications, New York

Marwala T (2010) Finite element model updating using computational intelligence techniques. Springer, London

Marwala T, Chakraverty S (2006) Fault classification in structures with incomplete measured data using autoassociative neural networks and genetic algorithm. Curr Sci 90:542–548

Marwala T, de Wilde P, Correia L, Mariano P, Ribeiro R, Abramov V, Szirbik N, Goossenaerts J (2001) Scalability and optimisation of a committee of agents using genetic algorithm. In: Proceedings of the international symposium on soft computing and intelligent systems for industry, Paisely, 2001

Michalewicz Z (1996) Genetic algorithms + data structures = evolution programs. Springer, New York

Mitchell M (1996) An introduction to genetic algorithms. MIT Press, Cambridge

Mohamed AK, Nelwamondo FV, Marwala T (2008) Estimation of missing data: neural networks, principal component analysis and genetic algorithms. In: Proceedings of the 12th world multiconference on systemics, cybernetics and informatics, Orlando, 2008, pp 36–41

Oh S, Pedrycz W (2006) Genetic optimization-driven multi-layer hybrid fuzzy neural networks. Simul Model Pract Theory 14:597–613

Oh KJ, Kim TY, Min S (2005) Using genetic algorithm to support portfolio optimization for index fund management. Expert Syst Appl 28:371–379

Park BJ, Choi HR, Kim HS (2003) A hybrid genetic algorithm for the job scheduling problems. Comput Ind Eng 45:597–613

Pendharkar PC, Rodger JA (1999) An empirical study of non-binary genetic algorithm-based neural approaches for classification. In: Proceedings of the 20th international conference on information systems, Chorlette, 1999, pp 155–165

Soleimani H, Golmakani HR, Salimi MH (2009) Markowitz-based portfolio selection with minimum transaction lots, cardinality constraints and regarding sector capitalization using genetic algorithm. Expert Syst Appl 36:5058–5063

Tettey T, Marwala T (2006) Controlling interstate conflict using neuro-fuzzy modeling and genetic algorithms. In: Proceedings of the 10th IEEE international conference on intelligent engineering systems, London, pp 30–44

Vose MD (1999) The simple genetic algorithm: foundations and theory. MIT Press, Cambridge

Chapter 10
Real-Time Approaches to Computational Economics: Self Adaptive Economic Systems

Abstract This chapter examines modelling of financial movement direction with Learn++ by forecasting the daily movement direction of the Dow Jones. The Learn++ approach is implemented using a multi-layer perceptron as a weak-learner, where this weak-learner is improved by making use of the Learn++ algorithm. In addition, the Learn++ algorithm introduces the concept of on-line incremental learning, which means that the proposed framework is able to adapt to new data.

10.1 Introduction

This chapter assumes that a complete model is the one that is able to continuously self-adapt to the changing environment. In this chapter, an on-line incremental algorithm that classifies the direction of movement of the stock market is described (Lunga and Marwala 2006a). One very important component of the economic system is the financial market. The financial market is a complex, evolving, and non-linear dynamic system. In order to increase the wealth of investors it is vital to be able to forecast the direction of the financial markets. The field of financial forecasting is manifested by data intensity, noise, non-stationarity, unstructured nature, high degree of uncertainty, and hidden relationships (Carpenter et al. 1992; Lunga and Marwala 2006a, b). Various aspects interact in finance, and these include social forces, political developments, overall economic conditions, and traders' expectations. Consequently, predicting market price movements is a difficult undertaking. Movements of market prices are not entirely random and they behave in a highly non-linear and dynamic manner. The standard random walk assumption of future prices is a different manifestation of randomness that hides a noisy non-linear process (McNelis 2005).

Incremental learning is a possible solution to such situations and is defined as the process of extracting new information from the data without losing prior knowledge from an additional dataset that later becomes available (Lunga and Marwala 2006a). A number of definitions and interpretations of incremental learning can be found in the literature, including on-line learning (Freund and Schapire 1997; Lunga and Marwala 2006b), re-learning of previously misclassified instances, and growing and pruning of classifier architectures (Bishop 1995). An algorithm possesses incremental learning capabilities if it meets the following criteria (Lunga and Marwala 2006b):

- Capability to attain further knowledge when new data are introduced.
- Capability to remember previously learned information about the data.
- Capability to learn new classes of data if introduced by new data.

Some applications of on-line classification problems have been reported recently (Polikar et al. 2002, 2004; Polikar 2000; Vilakazi et al. 2006; Vilakazi 2007). In many situations, the extent of accuracy and the acceptability of certain classifications are measured by the error of misclassified instances. Learn++ has mostly been applied to classification problems and the choice of Learn++ algorithm can boost a weak-learner to classify stock closing values with minimum error and reduced training time (Lunga and Marwala 2006b). For financial markets, forecasting methods based on minimizing forecasting error may not be sufficient. Trade driven by a certain forecast with a small forecast error may not be as profitable as trade guided by an accurate prediction of the direction of movement. This chapter discusses the ensemble systems, introduces the basic theory of incremental learning and the Learn++ algorithm, and applies these to financial markets.

10.2   Incremental Learning

An incremental learning algorithm is defined as an algorithm that learns new information from unseen data, without requiring access to previously observed data (Polikar et al. 2002, 2004; Polikar 2000). The algorithm is capable of learning newly available information from the data and to recall the knowledge from the previously observed data. Furthermore, the algorithm is capable of learning new classes that are introduced by subsequent data. This kind of learning algorithm is called an on-line learning procedure. Learning new information without accessing previously used data invokes the 'stability-plasticity dilemma' (Carpenter et al. 1992). A completely stable classifier retains the knowledge from previously learned data but fails to learn new information while a completely plastic classifier learns new data but forgets prior knowledge. The problem with neural network techniques is that they are stable classifiers and cannot learn new information after they have been trained. Different measures have been applied to capacitate neural networks with incremental learning capability. One technique of learning new information from supplementary data entails eliminating the trained classifier and training a new classifier using accumulated data. Other approaches such as pruning of networks or controlled alteration of classifier weight or growing of classifier architectures are known as incremental learning techniques and they change classifier weights using only the misclassified instances. These techniques are capable of learning new information, nevertheless, they suffer from 'forgetting' and necessitate access to old data. One method evaluates the current performance of the classifier architecture. If the present architecture does not adequately characterize the decision boundaries being learned, new decision clusters are generated in response to new pattern. Additionally, this method does not involve access to old data and can accommodate new classes. Nonetheless, the central inadequacies of this method are: cluster proliferation and sensitivity to selection of algorithm parameters. In this chapter, Learn++ is applied for on-line prediction of stock movement direction.

10.3 Ensemble Methods

The on-line learning technique implemented in this chapter is based on ensemble learning (Hansen and Salamon 1990; Jordan and Jacobs 1994; Kuncheva et al. 2001). *Ensemble learning* is a method where multiple models, such as classifiers, are deliberately created and combined to solve a particular problem (Rogova 1994; Polikar 2006; Marwala 2012). These techniques combine an ensemble of usually weak classifiers to exploit the so-called instability of the weak classifier (Polikar 2006). A tactical mixture of these classifiers eradicates the individual errors, creating a strong classifier. This makes the classifiers build adequately different decision boundaries for negligible changes in their training parameters and, as a result, each classifier makes different errors on any given instance. Ensemble systems have enticed a great deal of attention over the last decade due to their empirical success over single classifier systems on a variety of applications (Hulley and Marwala 2007; Marwala 2009).

Hannah and Dunson (2012) successfully applied an ensemble method to geo-metric programming based circuit design whereas Tong et al. (2012) successfully applied an ensemble of Kalman filters for approximating a heterogeneous con-ductivity field by integrating transient solute transport data. Austin et al. (2012) successfully applied ensemble methods for forecasting mortality in patients with cardiovascular disease whereas Halawani and Ahmad (2012) successfully applied ensemble methods to predict Parkinson disease and Ebrahimpour et al. (2012) successfully applied ensemble method to detect epileptic seizure.

In Sect. 10.3 ensemble learning methods are described: bagging, stacking and adaptive boosting (Marwala 2012). Particularly, the Adaptive Boosting technique is described because it was the basis for the creation of the Learn++ procedure, which is the on-line routine, implemented in this chapter (Polikar 2006).

10.3.1   Bagging

Bagging is a technique which is premised on the combination of models fitted to randomly chosen samples of a training data set to decrease the variance of the prediction model (Efron 1979; Breiman 1996; Marwala 2012). Bagging essentially necessitates randomly choosing a subset of the training data and applying this subset to train a model and repeating this process. Subsequently, all trained models are combined with equal weights to form an ensemble.

Louzada and Ara (2012) successfully applied bagging for fraud detection tool whereas Ghimire et al. (2012) successfully applied bagging for land-cover classification in Massachusetts. Syarif et al. (2012) successfully applied bagging to intrusion detection whereas Zhang et al. (2012) successfully applied bagging for high resolution range profile recognition for polarization radar.

10.3.2   Stacking

A model can be chosen from a set of models by comparing these models using data that was not used to train the models (Polikar 2006; Marwala 2012). This prior belief can also be applied to select a model amongst a set of models, based on a single data set by using a method called *cross-validation* (Bishop 1995; Marwala 2012). This is accomplished by dividing the data into a *training* data set, which is used to train the models, and a *testing* data set which is used to test the trained model. Stacking takes advantage of this prior belief by using the performance from the test data to combine the models instead of choosing among them the best performing model when tested on the testing data set (Wolpert 1992).

Sulzmann and Fürnkranz (2011) successfully applied stacking to compress an ensemble of rule sets into a single classifier whereas Chen and Wong (2011) successfully applied ant colony optimization method to optimize stacking ensemble. Lienemann et al. (2009) successfully applied stacking in metabonomic applications.

10.3.3   Adaptive Boosting (AdaBoost)

Boosting is a technique that incrementally generates an ensemble by training each new model with data that the previously trained model misclassified. Then the ensemble, which is a combination of all trained models, is used for prediction. Adaptive Boosting is an extension of boosting to multi-class problems (Freund and Schapire 1997; Schapire et al. 1998; Marwala 2012). There are many types of Adaptive Boosting, for instance AdaBoost.M1, where each classifier is assigned a weighted error of no more than ½ and AdaBoost.M2 with weak classifiers with a weighted error of less than ½. Tang et al. (2008) applied successfully Adaptive Boosting technique for analog circuit fault diagnosis whereas Li and Shen (2008) successfully applied Adaptive Boosting method for image processing. Nock et al. (2012) successfully applied boosting to classify natural scenes whereas Xia et al. (2012) successfully applied boosting for image retrieval. La et al. (2012) successfully applied boosting for text classification.

For AdaBoost.M1, samples are drawn from a distribution D that is updated in such a way that successive classifiers concentrate on difficult cases. This is achieved by adjusting D in such a way that the earlier, misclassified cases are likely to be present in the following sample. The classifiers are then combined through weighted majority voting. The distribution begins as a uniform distribution so that all cases have equal probability can be drawn into the first data subset S_1.

As described by Polikar (2006), at each iteration t, a new training data subset is sampled, and a weak classifier is trained to create a hypothesis h_t. The error given by this hypothesis with regards to the current distribution is estimated as the sum of distribution weights of the cases misclassified by h_t. AdaBoost.M1 requires that this error is less than ½, and if this requirement is violated then the procedure terminates. The normalized error β_t is then calculated so that the error that is in the [0 0.5] interval is normalized into the [0 1] interval. The transformed error is implemented in the distribution update rule, where $D_t(i)$ is decreased by a factor of $\beta_t, 0 < \beta_t < 1$, if x_i is correctly classified by h_t, or else it is left unaltered. When the distribution is normalized so that $D_{t+1}(i)$ is a proper distribution, the weights of those instances that are misclassified are increased. This update rule guarantees that the weights of all instances are correctly classified and the weights of all misclassified instances add up to ½. The requirement for the training error of the base classifier to be less than ½ forces the procedure to correct the error committed by the previous base model. When the training process is complete, the test data are classified by this ensemble of T classifiers, by applying a weighted majority voting procedure where each classifier obtains a voting weight that is inversely proportional to its normalized error (Polikar 2006). The weighted majority voting then selects the class ω allocated the majority vote of all classifiers. The procedure for Adaptive Boosting is shown in Algorithm 10.1 (Polikar 2006).

As described by Polikar (2006), the theoretical analysis of the Adaptive Boosting technique shows that the ensemble training error E is bounded above by:

$$E < 2^T \prod_{t=1}^{T} \sqrt{\varepsilon_t (1 - \varepsilon_t)} \tag{10.1}$$

The $\varepsilon_t < 1/2$ ensemble error E is reduced when new classifiers are added. The Adaptive Boosting method is not prone to over-fitting and this is explained by the margin theory (Schapire 1990; Polikar 2006).

Algorithm 10.1 The AdaBoost Algorithm.M1

Input:

- Input $X = \{x_1, x_2, \ldots, x_n\}$ and output $Y = \{y_1, y_2, \ldots, y_n\}$
- Weak-learner algorithm
- Number of classifiers T and distribution $D_1(i) = 1/n; i = 1, \ldots, n$

For $t = 1,2,\ldots,T$;

1. Sample a training subset S_t with a distribution D_t
2. Train *Weak-learner* with S_t and create hypothesis $h_t : X \rightarrow Y$
3. Estimate the error of h_t : $\varepsilon_t = \sum_{i=1}^{n} I[h_t(x_i) \neq y_i] \cdot D_t(i) = \sum_{t:h_t(x_i) \neq y_i} D_t(i)$
4. If $\varepsilon_t > \dfrac{1}{2}$ terminate
5. Estimate the normalized error $\beta_t = \varepsilon_t / (1 - \varepsilon_t) \Rightarrow 0 \leq \beta_t \leq 1$
6. Update the distribution D_t: $D_{t+1}(i) = \dfrac{D_t(i)}{Z_t} \times \begin{cases} \beta_t, \text{if } h_t(x_i) y_i \\ 1, otherwise \end{cases}$ where Z_t is the normalization constant so that D_{t+1} becomes a proper distribution function.

Test using majority voting given an unlabeled example z as follows:

- Count the total vote from the classifiers $V_j = \sum_{t:h_t(z)} \log(1/\beta_t)$ $j = 1, \ldots, C$
- Select the class that receives the highest number of votes as the final classification.

10.4   The Real-Time Method

Real-time learning is suitable for modelling dynamically time-varying systems where the characteristics of the environment in which the system is operating changes with time. It is also suitable when the data set existing is inadequate and does not entirely describe the system and, therefore, this approach incorporates new conditions that may be presented by newly acquired data.

A real-time computational economics model must have incremental learning competency if it is to be applied for automatic and continuous real-time prediction. The basis of real-time learning is incremental learning, which has been studied by many researchers (Higgins and Goodman 1991; Fu et al. 1996; Yamaguchi et al. 1999; Carpenter et al. 1992; Marwala 2012). The difficulty with real-time learning is the tendency of a real-time learner to forget the information learned during the initial stages of the learning process (McCloskey and Cohen 1989). The real-time learning technique adopted for this chapter is Learn++ and was proposed by Polikar et al. (2004).

Vilakazi and Marwala (2007a) applied the real-time incremental learning technique for monitoring the condition of high voltage bushings. Two incremental learning techniques were applied to the problem of condition monitoring. The first technique used was the incremental learning capability of the Fuzzy ARTMAP (FAM), and they investigated whether the ensemble approach can improve the performance of the FAM. The second technique applied was Learn++ that implemented an ensemble of multi-layer perceptron classifiers. Both methods performed well when tested for transformer bushing condition monitoring.

Mohamed et al. (2007) applied incremental learning for the classification of protein sequences. They used the fuzzy ARTMAP as an alternative machine learning system with the ability to incrementally learn new data as it becomes available. The fuzzy ARTMAP was seen to be comparable to many other machine learning systems. The application of an evolutionary strategy in the selection and combination of individual classifiers into an ensemble system, coupled with the incremental learning capability of the fuzzy ARTMAP was shown to be suitable as a pattern classifier. Their algorithm was tested using the data from the G-Coupled Protein Receptors Database and it demonstrated a good accuracy of 83 %.

Mohamed et al. (2006) applied fuzzy ARTMAP to multi-class protein sequence classification. They presented a classification system that used pattern recognition method to produce a numerical vector representation of a protein sequence and then classified the sequence into a number of given classes. They applied fuzzy ARTMAP classifiers and showed that, when coupled with a genetic algorithm based feature subset selection, the system could classify protein sequences with an accuracy of 93 %. This accuracy was then compared to other classification techniques and it was shown that the fuzzy ARTMAP was most suitable because of its high accuracy, quick training times and ability to learn incrementally.

Perez et al. (2010) applied a population-based, incremental learning approach to microarray gene expression feature selection. They evaluated the usefulness of the Population-Based Incremental Learning (PBIL) procedure on identifying a class differentiating gene set for sample classification. PBIL was based on iteratively evolving the genome of a search population by updating a probability vector, guided by the extent of class-separability demonstrated by a combination of features. The PBIL was then compared to standard Genetic Algorithm (GA) and an Analysis of Variance (ANOVA) method. The procedures were tested on a publicly available three-class leukemia microarray data set ($n = 72$). After running 30 repeats of both GA and PBIL, the PBIL could identify an average feature-space separability of 97.04 % while the GA achieved an average class-separability of 96.39 %. The PBIL also found smaller feature-spaces than GA, (PBIL – 326 genes and GA – 2,652) thus excluding a large percentage of redundant features. It also, on average, outperformed the ANOVA approach for $n = 2,652$ (91.62 %),

$q < 0.05$ (94.44 %), $q < 0.01$ (93.06 %) and $q < 0.005$ (95.83 %). The best PBIL run (98.61 %) even outperformed ANOVA for $n = 326$ and $q < 0.001$ (both 97.22 %). PBIL's performance was credited to its ability to direct the search, not only towards the optimal solution, but also away from the worst.

Hulley and Marwala (2007) applied GA-based incremental learning for optimal weight and classifier selection. They then compared Learn++, which is an incremental learning algorithm to the new Incremental Learning Using Genetic Algorithm (ILUGA). Learn++ demonstrated good incremental learning capabilities on benchmark datasets on which the new ILUGA technique was tested. ILUGA showed good incremental learning ability using only a few classifiers and did not suffer from catastrophic forgetting. The results obtained for ILUGA on the Optical Character Recognition (OCR) and Wine datasets were good, with an overall accuracy of 93 and 94 %, respectively, showing a 4 % improvement over Learn++. MT for the difficult multi-class OCR dataset.

Lunga and Marwala (2006a) applied a time series analysis using fractal theory and real-time ensemble classifiers to model the stock market. The fractal analysis was implemented as a concept to identify the degree of persistence and self-similarity within the stock market data. This concept was carried out using the Rescaled range analysis (R/S) technique. The R/S analysis outcome was then applied to a real-time incremental algorithm (Learn++) that was built to classify the direction of movement of the stock market. The use of fractal geometry in this study provided a way of determining, quantitatively, the extent to which the time series data could be predicted. In an extensive test, it was demonstrated that the R/S analysis provided a very sensitive technique to reveal hidden long runs and short run memory trends within the sample data. A time series data that was measured to be persistent was used to train the neural network. The results from the Learn++ algorithm showed a very high level of confidence for the neural network to classify sample data accurately.

Lunga and Marwala (2006b) applied incremental learning for the real-time forecasting of stock market movement direction. In particular, they presented a specific application of the Learn++ algorithm, and investigated the predictability of financial movement direction with Learn++ by forecasting the daily movement direction of the Dow Jones. The framework was implemented using the multi-layer perceptron (MLP) as a weak-learner. First, a weak learning algorithm, which tried to learn a class concept with a single input perceptron, was established. The Learn++ algorithm was then applied to improve the weak MLP learning capacity and thus introduced the concept of incremental real-time learning. The presented framework could adapt as new data were introduced and could classify the data well. This chapter is based on this study by Lunga and Marwala (2006b).

Vilakazi and Marwala (2007b) applied incremental learning to bushing condition monitoring. They presented a technique for bushing fault condition monitoring using the fuzzy ARTMAP. The fuzzy ARTMAP was introduced for bushing condition monitoring because it can incrementally learn information as it becomes available. An ensemble of classifiers was used to improve the classification accuracy of the system. The test results showed that the fuzzy ARTMAP ensemble gave an accuracy of 98.5 %. In addition, the results showed that the fuzzy ARTMAP could update its knowledge in an incremental fashion without forgetting the previously learned information.

Nelwamondo and Marwala (2007) successfully applied a technique for handling missing data from heteroskedasticity and non-stationary data. They presented a computational intelligence approach for predicting missing data in the presence of concept drift using an ensemble of multi-layer feed-forward neural networks. Six instances prior to the occurrence of missing data were used to approximate the missing values. The algorithm was applied to a simulated time series data set that resembled non-stationary data from a sensor. Results showed that the prediction of missing data in a non-stationary time series data was possible but was still a challenge. For one test, up to 78 % of the data could be predicted within a 10 % tolerance range of accuracy.

Khreich et al. (2012) conducted a survey of techniques for incremental learning of hidden Markov model parameters while Tscherepanow et al. (2011) applied hierarchical adaptive resonance theory network for the stable incremental learning of topological structures and associations from noisy data. Bouchachia (2011) studied incremental learning with multi-level adaptation. The author examined self-adaptation of classification systems which were natural adaptation of the base learners to change in the environment, contributive adaptation when combining the base learners in an ensemble, and structural adaptation of the combination as a form of dynamic ensemble. The author observed that this technique was able to deal with dynamic change in the presence of various types of data drift.

Martínez-Rego et al. (2011) proposed a robust incremental learning technique for non-stationary environments. They proposed a method, for single-layer neural networks, with a forgetting function in an incremental on-line learning procedure. The forgetting function offered a monotonically increasing significance to new data. Owing to the mixture of incremental learning and increasing significance assignment the network forgot quickly in the presence of changes while retaining a stable behavior when the context was stationary. The performance of the technique was tested over numerous regression and classification problems and the results were compared with those of previous works. The proposed procedure revealed high adaptation to changes while maintaining a low consumption of computational resources.

Yang et al. (2011) proposed an extreme and incremental learning based single-hidden-layer regularization ridgelet network which applied the ridgelet function as the activation function in a feed-forward neural network. The results showed that the method demonstrated incremental learning capability.

Topalov et al. (2011) successfully applied a neuro-fuzzy control of antilock braking system using a sliding mode incremental learning procedure. An incremental learning procedure was applied to update the parameters of the neuro-fuzzy controller. The application of this on the control of anti-lock breaking system model gave good results.

Folly (2011) proposed a method to optimally tune the parameters of power system stabilizers for a multi-machine power system using the Population-based incremental learning (PBIL) procedure. The PBIL procedure is a method that combines features of genetic algorithms and competitive learning-based on artificial neural networks. The results showed that PBIL based power system stabilizers performed better than genetic algorithm based power system stabilizers over a range of operating conditions considered.

Other successful implementations of incremental learning techniques include its use in anomaly detection (Khreich et al. 2009), in human robot interaction (Okada et al. 2009), for real-time handwriting recognition (Almaksour and Anquetil 2009), for predicting human movement in a vehicle motion (Vasquez et al. 2009), in visual learning (Huang et al. 2009), in nuclear transient identification (Baraldi et al. 2011), in object detection and pose classification (Tangruamsub et al. 2012), in classification of Alzheimer's disease (Cho et al. 2012), in face recognition (Lu et al. 2012) as well as in speech recognition (Li et al. 2012).

10.4.1   Learn++ Incremental Learning Method

Learn++ is an incremental learning procedure that was proposed by Polikar and co-workers (Polikar et al. 2002, 2004; Muhlbaier et al. 2004; Erdem et al. 2005; Polikar 2006; Marwala 2012). It is based on adaptive boosting procedure and applies multiple classifiers to capacitate the system to learn incrementally. The procedure operates on the notion of using many classifiers that are weak-learners to give a good overall classification. The weak-learners are trained on a separate subset of the training data and then the classifiers are combined using a weighted majority vote. The weights for the weighted majority vote are selected using the performance of the classifiers on the entire training dataset.

Each classifier is trained using a training subset that is sampled in accordance to a stated distribution. The classifiers are trained using a weak-learner approach. The condition for the weak-learner procedure is that it must give a classification rate of less than 50 % firstly (Polikar et al. 2002). For each database D_k that contains training series, S, where S contains learning examples and their equivalent classes, Learn++ starts by setting the weights vector, w, according to a specified distribution D_T, where T is the number of hypothesis. Firstly the weights are set to be uniform giving equal probability for all cases chosen for the first training subset and the distribution is then given by (Polikar et al. 2002; Marwala 2012):

$$D = \frac{1}{m} \tag{10.2}$$

Here, m is the number of training examples in S. The training data are then distributed into training subset TR and testing subset TE to confirm the weak-learner capability. The distribution is then applied to choose the training subset TR and testing subset TE from S_k. After training and testing subsets have been chosen, then the weak-learner procedure is applied. The weak-learner is trained using subset *TR*. A hypothesis, h_t, attained from a weak-learner is tested using both the training and testing subsets to achieve an error (Polikar et al. 2002; Marwala 2012):

$$\varepsilon_t = \sum_{t:h_i(x_i)\neq y_i} D_t(i) \tag{10.3}$$

The error is required to be less than 0.5; a normalized error is computed using (Polikar et al. 2002; Marwala 2012):

$$B_t = \frac{\varepsilon_t}{1-\varepsilon_t} \tag{10.4}$$

If the error is greater than 0.5, the hypothesis is rejected and the new training and testing subsets are chosen according to a distribution D_T and another hypothesis is estimated. All classifiers created are then combined using weighted majority voting to obtain a combined hypothesis, H_t (Polikar et al. 2002; Marwala 2012):

$$H_t = \arg\max_{y\in Y} \sum_{t:h_t(x)=y} \log\left(\frac{1}{\beta_t}\right) \tag{10.5}$$

The weighted majority voting offers higher voting weights to a hypothesis that performs well on the training and testing data subsets. The error of the composite hypothesis is calculated as follows (Polikar et al. 2002; Marwala 2012):

$$E_t = \sum_{t:H_i(x_i)\neq y_i} D_t(i) \tag{10.6}$$

If the error is greater than 0.5, the current hypothesis is rejected and the new training and testing data are chosen according to a distribution D_T. Or else, if the error is less than 0.5, then the normalized error of the composite hypothesis is calculated as follows (Polikar et al. 2002; Marwala 2012):

$$B_t = \frac{E_t}{1-E_t} \tag{10.7}$$

The error is applied in the distribution update rule, where the weights of the correctly classified cases are reduced, accordingly increasing the weights of the mis-classified instances. This confirms that the cases that were misclassified by the current hypothesis have a higher probability of being chosen for the succeeding training set. The distribution update rule is given by the following equation (Polikar et al. 2002; Marwala 2012):

$$w_{t+1} = w_t(i) \times B_t^{1-[|H_t(x_i)\neq y_i|]} \tag{10.8}$$

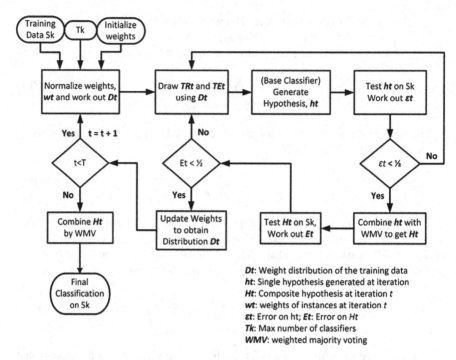

Fig. 10.1 Learn++ algorithm

After the T hypothesis has been generated for each database, the final hypothesis is calculated by combining the hypotheses using weighted majority voting as described by the following equation (Polikar et al. 2002; Marwala 2012):

$$ H_t = \arg\max_{y \in Y} \sum_{k=1}^{K} \sum_{t:H_t(x)=y} \log\left(\frac{1}{\beta_t}\right) \tag{10.9} $$

The Learn++ algorithm is represented diagrammatically in Fig. 10.1 (Polikar 2006; Marwala 2012).

10.4.2 Confidence Measurement

To approximate the confidence of the Learn++ procedure, a majority of hypotheses agreeing on given instances is an indicator of confidence on the decision proposed. If it is assumed that a total of T hypotheses are generated in k training sessions for a C-class problem, then for any given example, the final classification class, the total vote class c received is given by (Muhlbaier et al. 2004; Marwala 2012):

Table 10.1 Confidence
estimation representation
(Lunga and Marwala 2006b)

Confidence range (%)	Confidence level
$0.9 \leq \lambda c \leq 1$	Very high (VH)
$0.8 \leq \lambda c \leq 0.8$	High (H)
$0.7 \leq \lambda c \leq 0.8$	Medium (M)
$0.6 \leq \lambda c \leq 0.7$	Low (l)
$0 \leq \lambda c \leq 0.6$	Very low (VL)

$$\zeta_c = \sum_{t:h_t(x)=c} \Psi_t \qquad (10.10)$$

where Ψ_t denotes the voting weights of the tth, hypothesis h_t.

Normalizing the votes received by each class can be performed as follows (Muhlbaier et al. 2004; Marwala 2012):

$$\lambda_c = \frac{\zeta_c}{\sum\limits_{c=1}^{C} \zeta_c} \qquad (10.11)$$

Here, λ_c can be interpreted as a measure of confidence on a scale of 0–1 and this representation is shown in Table 10.1 (Lunga and Marwala 2006b). A high value of λ_c shows high confidence in the decision and conversely, a low value of λ_c shows low confidence in the decision. It should be noted that the λ_c value does not represent the accuracy of the results, but the confidence of the system in its own decision.

10.4.3 Multi-layer Perceptron

In this chapter we use the multi-layer perceptron neural network to create a weak-learner. The multi-layered perceptrons have been successfully used to model complex systems (Marwala 2007), missing data estimation (Marwala 2009), inter-state conflict modelling (Marwala and Lagazio 2011) and condition monitoring (Marwala 2012). Each connection between inputs and neurons is weighted by adjustable weight parameters. Furthermore, each neuron has an adjustable bias weight parameter which is represented by a connection from a constant input $x_0 = 1$ and $z_0 = 1$ for the hidden neurons and the output neuron, respectively. This group of two-layer multi-layer perceptron models is capable of estimating any continuous function with arbitrary accuracy, providing the number of hidden neurons is appropriately large (Bishop 1995).

The advantage of the multi-layer perceptron network is the interconnected cross-coupling that occurs between the input variables and the hidden nodes, and the hidden nodes and the output variables. If we assume that x is the input to the multi-layer perceptron and y is the output of the MLP, a mapping function between the input and the output may be written as follows (Bishop 1995):

$$y = f_{\text{output}} \left(\sum_{j=1}^{M} w_j \, f_{\text{hidden}} \left(\sum_{i=0}^{N} w_{ij} x_i \right) + w_0 \right) \qquad (10.12)$$

where N is the number of input units, M is the number of hidden neurons, x_i is the ith input unit, w_{ij} is the weight parameter between input i and hidden neuron j and w_j is the weight parameter between hidden neuron j and the output neuron. The activation function $f_{\text{output}}(\cdot)$ is sigmoid and can be written as follows (Bishop 1995):

$$f_{output}(a) = \frac{1}{1 + e^{-a}} \qquad (10.13)$$

For classification problems, the sigmoid function is ideal (Bishop 1995). The activation function $f_{\text{hidden}}(\cdot)$ is a hyperbolic tangent can be written as follows (Bishop 1995):

$$f_{hidden}(a) = \tanh(a) \qquad (10.14)$$

The neural network model in Eq. 10.12 is trained using the scaled conjugate gradient method, which is described in Bishop (1995).

10.5   Experimental Investigation

This analysis examines the daily changes of the Dow Jones Index. The Dow Jones averages are particular in that they are price weighted rather than market capitalization weighted. Their component weightings are consequently impacted only by changes in the stock prices, in contrast with other indexes' weightings that are impacted by both price changes and changes in the number of shares outstanding (Leung et al. 2000). When the averages were originally generated, their values were computed by merely totalling up the constituent stock prices and dividing by the number of constituents. Altering the divisor was started to isolate the consequences of stock separations and other corporate activities.

The Dow Jones Industrial Average measures the composite price performance of over 30 highly capitalized stocks trading on the New York Stock Exchange (NYSE), representing a broad cross-section of industries in the USA. Trading in the index has gained unparalleled reputation in foremost financial markets around the world. The increasing diversity of financial instruments associated to the Dow Jones Index has expanded the dimension of global investment prospect for both individual and institutional investors. There are two basic explanations for the success of these

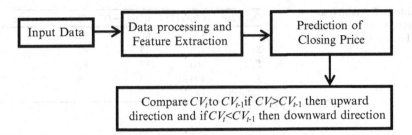

Fig. 10.2 Proposed model for real-time stock forecasting

index trading instruments. The first reason is that they afford an effective means for investors to hedge against potential market risks. The second reason is that they generate new profit making prospects for market investors. Consequently, it has deep consequences and importance for researchers and practitioners to correctly forecast the direction of the movement of stock prices.

Previous research has investigated the cross-sectional relationship between stock index and macroeconomic variables. Macroeconomic input variables which are normally implemented for forecasting include term structure of interest rates, short-term interest rate, long-term interest rate, consumer price index, industrial production, government consumption, private consumption and gross domestic product. In this chapter, the closing values of the index were selected as inputs.

A one step forward prediction of the index was performed on a daily basis. The output of this prediction model was used as input to the Learn++ algorithm for classification into the correct category that would give an indication of whether the predicted index value is 1 (indicating a positive increase in next day's predicted closing value compared to the previous day's closing value) or a predicted closing value of -1, indicating a decrease in next day's predicted closing value compared to the previous day's closing value. Figure 10.2 shows the conceptual model of all processes needed for this study (Lunga and Marwala 2006a). The first prediction model can be written as (Lunga and Marwala 2006a):

$$CV_t = F\left(cv_{t-1}, cv_{t-1}, cv_{t-1}, cv_{t-1}\right) \tag{10.15}$$

where CV_t is the predicted close value at time t, cv_{t-1} indicates the close value at day i, where $i = 1,2,3, t-1$. The second model takes the output of the first model as its input in predicting the direction of movement for the index. The classification prediction stage can be represented as (Lunga and Marwala 2006a):

$$Direction_t = F\left(CV_t\right) \tag{10.16}$$

where CV_t is the first model prediction of the fifth day stock closing value when given the raw data at time $t-1$ to $t-4$ respectively. $Direction_t$ is a categorical

Table 10.2 Training and generalisation performance of Learn++

Database	Class (1)	Class (−1)	Test performance (%)
S_1	132	68	72
S_2	125	75	82
S_3	163	37	85
S_4	104	96	86
Validate	143	57	−

variable to indicate the movement direction of the Dow Jones Index at time t. If the Dow Jones Index at time t is larger than that at time $t − 1$, *Direction* is 1, otherwise, *Direction* is −1.

The model estimation selection process is then followed by an empirical evaluation which is based on the out-of-sample data. At this step, the comparative performance of the model is measured by the classification accuracy of the final hypothesis chosen for all given databases. The confidence of the algorithm on its own decision is used to evaluate the accuracy of the predicted closing value category. The first experiment implements a one step forward prediction of the next day's stock closing value. After predicting the next day's closing value this value is fed into a classification model to indicate the direction of movement for the stock prices. As discussed above, the database consisted of 1,476 instances of the Dow Jones average closing value during the period from January 2000 to November 2005; 1,000 instances are used for training and all the remaining instances are used for validation (Lunga and Marwala 2006a). The two binary classes are 1, indicating an upward direction of returns in Dow Jones stock, and −1 to indicate a predicted fall/downward direction of movement for the Dow Jones stock.

Four datasets S_1, S_2, S_3 and S_4, where each dataset included exactly one quarter of the entire training data, were provided to Learn++ in four training sessions for incremental learning. For each training session k ($k = 1, 2, 3, 4$) three weak hypothesis were produced by Learn++. Each hypothesis h_1, h_2, and h_3 of the kth training session was produced using a training subset TR_t and a testing subset TE_t. The weak-learner was a single hidden layer multi-layer perceptron with 15 hidden layer nodes and 1 output node with an MSE goal of 0.1. The testing set of data consisted of 476 instances that were used for validation purposes. On average, the multi-layer perceptron hypothesis, weak-learner, performed little over 50 %, which improved to over 80 % when the hypotheses were combined by making use of weighted majority voting. This improvement demonstrated the performance improvement property of Learn++, as inherited from Adaptive Boosting, on a given database. The data distribution and the percentage classification performance are given in Table 10.2 (Lunga and Marwala 2006b). The performances listed are on the validation data.

Table 10.3 gives an actual breakdown of correctly classified and misclassified instances falling into each confidence range after each training session. The trends of the confidence estimates after subsequent training sessions are given in Table 10.3. The desired outcome on the actual confidences is high to very high

Table 10.3 Confidence results

		VH	H	M	VL	L
Correctly classified	S_1	96	96	13	15	6
	S_2	104	104	22	17	14
	S_3	111	111	6	3	39
	S_4	101	101	42	12	4
Incorrectly classified	S_1	23	7	13	3	8
	S_2	27	0	1	3	4
	S_3	21	1	2	4	2
	S_4	24	0	2	2	0

Table 10.4 Confidence trends for Dow Jones

	Increasing steady	Decreasing
Correctly classified	119	8
Misclassified	16	24

confidences on correctly classified instances, and low to very low confidences on misclassified instances. The desired outcome on confidence trends is increasing or steady confidences on correctly classified instances, and decreasing confidences on misclassified instances, as new data is introduced.

The performance shown in Table 10.2 indicates that the algorithm is improving its generalization capacity as new data become available. The improvement is modest, however, as majority of the new information is already learned in the first training session. Table 10.4 indicates that the vast majority of correctly classified instances tend to have very high confidences, with continually improved confidences at consecutive training sessions (Lunga and Marwala 2006a).

While a considerable portion of misclassified instances also had high confidence for this database, the general desired trends of increased confidence on correctly classified instances and decreasing confidence on misclassified ones were notable and dominant, as shown in Table 10.3 (Lunga and Marwala 2006a).

10.6 Conclusions

In this chapter, an incremental learning procedure, Learn++, was applied to predict the financial markets movement direction. Learn++ is found to provide good results on adapting the weak-learner (MLP) into a strong learning algorithm that has confidence in all its decisions. The Learn++ procedure was found to evaluate the confidence of its own decisions. Generally, the majority of correctly classified cases had very high confidence approximations while lower confidence values were related with misclassified cases. Consequently, classification cases with low confidences can be further evaluated. In addition, the procedure also demonstrated increasing confidences in correctly classified instances and decreasing confidences in misclassified instances after successive training sessions.

References

Almaksour A, Anquetil E (2009) Fast incremental learning strategy driven by confusion reject for on-line handwriting recognition. In: Proceedings of the international conference on document analysis and recognition, Barcelona, 2009, pp 81–85

Austin PC, Lee DS, Steyerberg EW, Tu JV (2012) Regression trees for predicting mortality in patients with cardiovascular disease: what improvement is achieved by using ensemble-based methods? Biom J 54:657–673

Baraldi P, Razavi-Far R, Zio E (2011) Classifier-ensemble incremental-learning procedure for nuclear transient identification at different operational conditions. Reliab Eng Syst Saf 96: 480–488

Bishop C (1995) Neural networks for pattern recognition. Oxford University Press, Oxford

Bouchachia A (2011) Incremental learning with multi-level adaptation. Neurocomputing 74:1785–1799

Breiman L (1996) Bagging predictors. Mach Learn 24:123–140

Carpenter G, Grossberg S, Marhuzon N, Reynolds J, Rosen D (1992) ARTMAP: a neural network architecture for incremental learning supervised learning of analog multi-dimensional maps. IEEE Trans Neural Netw 3:678–713

Chen Y, Wong ML (2011) Optimizing stacking ensemble by an ant colony optimization approach. In: Proceedings of the genetic and evolutionary computation conference, Dublin, 2011, pp 7–8

Cho Y, Seong J-K, Jeong Y, Shin SY (2012) Individual subject classification for Alzheimer's disease based on incremental learning using a spatial frequency representation of cortical thickness data. Neuroimage 59:2217–2230

Ebrahimpour R, Babakhani K, AbbaszadehArani SAA, Masoudnia S (2012) Epileptic seizure detection using a neural network ensemble method and wavelet transform. Neural Netw World 22:291–310

Efron B (1979) Bootstrap methods: another look at the jackknife. Ann Stat 7:1–26

Erdem Z, Polikar R, Gurgen F, Yumusak N (2005) Reducing the effect of out-voting problem in ensemble based incremental support vector machines. Lect Note Comput Sci 3697:607–612

Folly KA (2011) Performance evaluation of power system stabilizers based on population-based incremental learning (PBIL) algorithm. Int J Electr Power Energy Syst 33:1279–1287

Freund Y, Schapire R (1997) A decision-theoretic generalization of on-line learning and an application to boosting. J Comput Syst Sci 55:119–139

Fu L, Hsu HH, Principe JC (1996) Incremental backpropagation networks. IEEE Trans Neural Netw 7:757–761

Ghimire B, Rogan J, Galiano V, Panday P, Neeti N (2012) An evaluation of bagging, boosting, and random forests for land-cover classification in Cape Cod, Massachusetts, USA. GISci Remote Sens 49:623–643

Halawani SM, Ahmad A (2012) Ensemble methods for prediction of Parkinson disease. Lect Note Comput Sci 7435:516–521

Hannah LA, Dunson DB (2012) Ensemble methods for convex regression with applications to geometric programming based circuit design. In: Proceedings of the 29th international conference on machine learning, Edinburgh, 2012, pp 369–376

Hansen LK, Salamon P (1990) Neural network ensembles. IEEE Trans Pattern Anal Mach Intell 12:993–1001

Higgins CH, Goodman RM (1991) Incremental learning for rule based neural network. In: Proceedings of the international joint conference on neural networks, Seattle, 1991, pp 875–880

Huang D, Yi Z, Pu X (2009) A new incremental PCA algorithm with application to visual learning and recognition. Neural Process Lett 30:171–185

Hulley G, Marwala T (2007) Genetic algorithm based incremental learning for optimal weight and classifier selection. In: Proceedings of the AIP conference, Sydney, Australia, pp 258–267

Jordan MJ, Jacobs RA (1994) Hierarchical mixtures of experts and the EM algorithm. Neural Comput 6:181–214

Khreich W, Granger E, Miri A, Sabourin RA (2009) A comparison of techniques for on-line incremental learning of HMM parameters in anomaly detection. In: Proceedings of the IEEE symposium on computational intelligence for security and defense applications, Ottawa,2009, pp 1–8

Khreich W, Granger E, Miri A, Sabourin R (2012) A survey of techniques for incremental learning of HMM parameters. Inf Sci 197:105–130

Kuncheva LI, Bezdek JC, Duin R (2001) Decision templates for multiple classifier fusion: an experimental comparison. Pattern Recognit 34:299–314

La L, Guo Q, Yang D, Cao Q (2012) Multiclass boosting with adaptive group-based k-NN and its application in text categorization. Math Probl Eng 2012:1–24

Leung M, Daouk H, Chen A (2000) Forecasting stock indices: a comparison of classification and level estimation models. Int J Forecast 16:173–190

Li H, Shen C (2008) Boosting the minimum margin: LP boost vs. ada boost. In: Proceedings of the digital image computing: techniques and applications, Canberra, Australia, pp 533–539

Li H, Zhang T, Qiu R, Ma L (2012) Grammar-based semi-supervised incremental learning in automatic speech recognition and labeling. Energy Procedia 17:1843–1849

Lienemann K, Plötz T, Fink GA (2009) Stacking for ensembles of local experts in metabonomic applications. Lect Note Comput Sci 5519:498–508

Louzada F, Ara A (2012) Bagging k-dependence probabilistic networks: an alternative powerful fraud detection tool. Expert Syst Appl 39:11583–11592

Lu G-F, Zou J, Wang Y (2012) Incremental learning of complete linear discriminant analysis for face recognition. Knowl-Based Syst 31:19–27

Lunga D, Marwala T (2006a) Time series analysis using fractal theory and on-line ensemble classifiers. Lect Note Comput Sci 4304:312–321

Lunga D, Marwala T (2006b) On-line forecasting of stock market movement direction using the improved incremental algorithm. Lect Note Comput Sci 4234:440–449

Martínez-Rego D, Pérez-Sánchez B, Fontenla-Romero O, Alonso-Betanzos A (2011) A robust incremental learning method for non-stationary environments. Neurocomputing 74:1800–1808

Marwala T (2007) Computational intelligence for modelling complex systems. Research India Publications, New Delhi

Marwala T (2009) Computational intelligence for missing data imputation, estimation and management: knowledge optimization techniques. IGI Global Publications, New York

Marwala T (2012) Condition monitoring using computational intelligence methods. Springer, London

Marwala T, Lagazio M (2011) Militarized conflict modeling using computational intelligence techniques. Springer, London

McCloskey M, Cohen N (1989) Catastrophic interference connectionist networks: the sequential learning problem. Psychol Learn Motiv 24:109–164

McNelis PD (2005) Neural networks in finance: gaining the predictive edge in the market. Elsevier Academic Press, Oxford

Mohamed S, Rubin D, Marwala T (2006) Multi-class protein sequence classification using fuzzy ARTMAP. In: Proceedings of the IEEE international conference on systems, man, and cybernetics, Taipei, 2006, pp 1676–1681

Mohamed S, Rubin D, Marwala T (2007) Incremental learning for classification of protein sequences. In: Proceedings of the IEEE international joint conference on neural networks, Orlando, 2007, pp 19–24

Muhlbaier M, Topalis A, Polikar R (2004) Learn++.MT: a new approach to incremental learning. In: Proceedings of the 5th international workshop on multiple classifier systems, Cagliari, 2004, pp 52–61

Nelwamondo FV, Marwala T (2007) Handling missing data from heteroskedastic and nonstationary data. Lect Note Comput Sci 4491:1293–1302

Nock R, Piro P, Nielsen F, Bel Haj Ali W, Barlaud M (2012) Boosting k-NN for categorization of natural scenes. Int J Comput Vis 100:294–314

Okada S, Kobayashi Y, Ishibashi S, Nishida T (2009) Incremental learning of gestures for human-robot interaction. AI Soc 25:155–168

Perez M, Featherston J, Marwala T, Scott LE, Stevens DM (2010) A population-based incremental learning approach to microarray gene expression feature selection. In: Proceedings of the IEEE 26th convention of electrical and electronic engineers, Eilat, 2010, pp 10–14

Polikar R (2000) Algorithms for enhancing pattern separability, feature selection and incremental learning with applications to gas sensing electronic noise systems. Ph.D. thesis, Iowa State University, Ames

Polikar R (2006) Ensemble based systems in decision making. IEEE Circuit Syst Mag 6:21–45

Polikar R, Byorick J, Krause S, Marino A, Moreton M (2002) Learn++: a classifier independent incremental learning algorithm for supervised neural network. Proc Int Jt Conf Neural Netw 2:1742–1747

Polikar R, Udpa L, Udpa S, Honavar V (2004) An incremental learning algorithm with confidence estimation for automated identification of NDE signals. Trans Ultrason Ferroelectr Freq Control 51:990–1001

Rogova G (1994) Combining the results of several neural network classifiers. Neural Netw 7:777–781

Schapire RE (1990) The strength of weak learnability. Mach Learn 5:197–227

Schapire RE, Freund Y, Bartlett P, Lee WS (1998) Boosting the margin: a new explanation for the effectiveness of voting methods. Ann Stat 26:51–1686

Sulzmann J-N, Fürnkranz J (2011) Rule stacking: an approach for compressing an ensemble of rule sets into a single classifier. Lect Note Comput Sci 6926:323–334

Syarif I, Zaluska E, Prugel-Bennett A, Wills G (2012) Application of bagging, boosting and stacking to intrusion detection. Lect Note Comput Sci 7376:593–602

Tang J, Shi Y, Zhou L, Zhang W (2008) Analog circuit fault diagnosis using Ada boost and SVM. In: Proceedings of the international conference on communications, circuits and systems, Fujian, China, pp 1184–1187

Tangruamsub S, Takada K, Hasegawa O (2012) A fast on-line incremental learning method for object detection and pose classification using voting and combined appearance modelling. Signal Process Image Commun 27:75–82

Tong J, Hu BX, Yang J (2012) Data assimilation methods for estimating a heterogeneous conductivity field by assimilating transient solute transport data via ensemble Kalman filter. Hydrol Process. doi:10.1002/hyp.9523

Topalov AV, Oniz Y, Kayacan E, Kaynak O (2011) Neuro-fuzzy control of antilock braking system using sliding mode incremental learning algorithm. Neurocomputing 74:1883–1893

Tscherepanow M, Kortkamp M, Kammer M (2011) A hierarchical ART network for the stable incremental learning of topological structures and associations from noisy data. Neural Netw 24:906–916

Vasquez D, Fraichard T, Laugier C (2009) Growing hidden markov models: an incremental tool for learning and predicting human and vehicle motion. Int J Robot Res 28:1486–1506

Vilakazi B (2007) Machine condition monitoring using artificial intelligence: the incremental learning and multi-agent system approach. M.Sc. thesis, University of the Witwatersrand, Johannesburg

Vilakazi CB, Marwala T (2007a) Incremental learning and its application to bushing condition monitoring. Lect Note Comput Sci 4491:1237–1246

Vilakazi CB, Marwala T (2007b) On-line incremental learning for high voltage bushing condition monitoring. In: Proceedings of the international joint conference on neural networks, Orlando, 2007, pp 2521–2526

Vilakazi B, Marwala T, Mautla R, Moloto E (2006) On-line bushing condition monitoring using computational intelligence. WSEAS Trans Power Syst 1:280–287

Wolpert DH (1992) Stacked generalization. Neural Netw 5:241–259

Xia H, Wu P, Hoi SCH, Jin R (2012) Boosting multi-kernel locality-sensitive hashing for scalable image retrieval. In: Proceedings of the international ACM SIGIR conference on research and development in information retrieval, Portland, Oregon, pp 55–64

Yamaguchi K, Yamaguchi N, Ishii N (1999) Incremental learning method with retrieving of interfered patterns. IEEE Trans Neural Netw 10:1351–1365

Yang S, Wang M, Jiao L (2011) Extreme and incremental learning based single-hidden-layer regularization ridgelet network. Neurocomputing 74:1809–1814

Zhang Y-X, Wang X-D, Yao X, Bi K (2012) HRRP recognition for polarization radar based on bagging-SVM dynamic ensemble. Syst Eng Electron 34:1366–1371

Chapter 11
Multi-agent Approaches to Economic Modeling: Game Theory, Ensembles, Evolution and the Stock Market

Abstract A multi-agent system that learns by using neural networks is implemented to simulate the stock market. Each committee of agents, which is regarded as a player in a game, is optimized by continually adapting the architecture of the agents through the use of genetic algorithms. The proposed procedure is implemented to simulate trading of three stocks, namely, the Dow Jones, the NASDAQ and the S&P 500.

11.1 Introduction

In this chapter a committee of agents is used to simulate the stock market (Marwala et al. 2001). Each committee of agents is viewed as a player in a game and, therefore, a game theoretic framework is applied in this chapter (Marwala et al. 2001). These players in a game compete and cooperate (Perrone and Cooper 1993). The committee of agents is optimized using a genetic algorithm (Holland 1975; Goldberg 1989). Perrone and Cooper (1993) introduced a committee of networks, which optimizes the decision-making of a population of non-linear predictive models (Bishop 1995). They attained this by assuming that the trained predictive models were accessible and then allocating, to each network, a weighting factor, which specifies the role that the network has on the total decision of a population of networks. The drawback of their proposal is that, in a condition where the problem is altering such as the stock market, the technique is not sufficiently elastic to permit for the dynamic evolution of the population of networks.

This chapter aims to relax this constraint on the committee technique by ensuring that the individual networks that create the committee are permitted to dynamically evolve as the problem evolves using genetic programming (Michalewicz 1996), and this was first conducted by Marwala et al. (2001).

The parameters describing the design of the networks, such as the number of hidden units, which form a committee, are defined as design variables and are permitted to evolve as the trading environment evolves. The network characteristics that are appropriate for survival substitute for those that are not appropriate. On applying a genetic algorithm to choose the appropriate individuals, three steps are followed: (1) crossover of network attributes within the population; (2) Mutation of each individual attributes; and (3) reproduction of the successful attributes. The simple crossover, the binary mutation, and roulette wheel reproduction techniques are used.

In conclusion, the proposed technique is applied to simulate the trading of three stocks. The scalability of the number of agents and players in the simulations with respect to computational time were investigated. The evolution of the complexity of the simulation as the players participate in more trading was also investigated. The next section describes game theory, which is a framework that is used to set up the simulation.

11.2   Game Theory

In this chapter, we apply game theory to model the stock market. Game theory essentially consists of players, set of actions (strategy), and pay-off function (Villena and Villena 2004; Ross 2006; van den Brink et al. 2008). Game theory has been applied to many areas of activity including economics (van den Brink et al. 2008), procurement of land (Hui and Bao 2013), auction (Laffont 1997), the hotel industry (Wei et al. 2012), facial recognition (Roy and Kamel 2012), medicine (McFadden et al. 2012) and computer science (Papadimitriou 2001). There are many types of games and, in this chapter we will illustrate the well-known prisoner's dilemma problem. Suppose players A and B are arrested for a crime and they are put into separate cells. They are given choices to either cooperate or defect, and this is represented in Table 11.1.

Game theory can be used to solve the problem in Table 11.1. In this table, if a player remains slilent, he gets either 2 months in prison or serves 1 year in prison. If he pleas bargains, he gets either 6 months in prison or goes free. According to John von Neumann, the best strategy is the one that guarantees you maximum possible outcome even if your opponent knew what choice you were going to make. In this case, the best strategy is to enter a plea bargain. The concept of Nash equilibrium states that the best strategy for each player is such that every player's move is a best

Table 11.1 Illustration of the prisoner's dilemma

	Prisoner B remains silent	Prisoner B plea bargains
Prisoner A remains silent	Each serves 2 months	Prisoner A serves 1 year Prisoner B goes free
Prisoner A plea bargains	Prisoner B serves 1 year Prisoner A goes free	Each serves 6 months

response to the other players' move. Therefore, entering a plea bargain is a Nash equilibrium. Of course, all these assume that each player is rational and aims to maximize pay-off (Beed and Beed 1999).

Hodgson and Huang (2012) compared evolutionary game theory and evolutionary economics and concluded that these methods improve understanding of structures and causal processes, whereas Christin (2011) applied game theory in network security games and concluded that it was vital to understand reasons of different players in a network to design systems to support appropriate outcome.

Hanauske et al. (2010) extended the hawk-dove game by a quantum method and demonstrated that evolutionary stable strategies, which are not forecast by traditional evolutionary game theory and where the total economic population applies a non-aggressive quantum strategy, can also emerge.

McCain (2009) studied theoretical and experimental results in game theory and, the neo-classical notion of inter-temporal discrepancy in choice to debate that the motivational theory which is common between neo-classical economics and non-cooperative game theory, mistakenly assumes that commitment never occurs in human decisions. They concluded that the conditions that favor commitment, other than alterations of an assumed utility, function to account for non-self-regarding motivations are advantageous in behavioral economics and game theory.

Roth (2002) applied game theory to the design of the entry level labor market for American doctors and the auctions of radio spectrum. He proposed that experimental and computational economics complemented game theory for design and debated that some of the tasks confronting both markets include handling with associated types of complementarities.

The example illustrated at the beginning of this chapter was a two player game. It becomes extremely difficult to deal with multiple player games and a computational technique has been developed and is able to handle, to some extent, multiple player games and this procedure is called a multi-agent system and is the subject of the next section.

11.3 Multi-agent Systems

A multi-agent system is, by definition, a system of multiple agents. An agent is an object that is autonomous, perceives its environment, and acts on its environment is intelligent, and operates autonomously in that environment (Russell and Norvig 2003; Franklin and Graesser 1996; Kasabov 1998). Agents have the following characteristics (Kasabov 1998):

- They are autonomous.
- They are flexible, reactive, proactive and social.
- They have control capability.

To illustrate the working of a multi-agent system, a well-known swarm intelligence theory can be used. In this example, agents or birds (in the case of the swarming of birds) operate using two simple rules. These are that, in seeking the next move, a bird considers the best position it has encountered and the best position the entire flock has encountered (where other birds are going). Using these simple rules, the swarm is able to solve very complex problems. More details on these can be found in the literature (Marwala 2009, 2010, 2012; Marwala and Lagazio 2011).

Teweldemedhin et al. (2004) presented an agent-based, bottom-up modeling technique to develop a simulation tool for estimating and predicting the spread of the human immunodeficiency virus (HIV) in a given population. They developed a simulation instrument to understand the spread of HIV.

Hurwitz and Marwala (2007) studied the deed of bluffing, which has perplexed game designers. They asserted that, the very act of bluffing was even open for debate, introducing additional difficulty to the procedure of producing intelligent virtual players that can bluff, and therefore play, truthfully. Through the application of intelligent, learning agents, and prudently designing agents, an agent was found to learn to predict its opponents' reactions based on its own cards and actions of other agents. They observed that, an agent can learn to bluff its opponents, with the action not indicating an "irrational" action as bluffing is usually regarded, but as an act of maximizing returns by an actual statistical optimization. They applied a TD lambda learning algorithm to adapt a neural network based agent's intelligence and demonstrated that agents were able to learn to bluff without outside encouragement.

Abdoos et al. (2011) applied a multi agent technique for traffic light control in non-stationary environments. The results they obtained indicated that the proposed method performed better than a fixed time technique under different traffic demands. Elammari and Issa (2013) applied model driven architecture to develop multi-agent systems, while Chitsaz and Seng (2013) successfully applied a multi agent system for medical image segmentation.

Stroeve et al. (2013) successfully applied event sequence analysis and multi-agent systems for safety assessments of a runway incursion scenario, while El-Menshawy et al. (2013) verified, successfully, conformance of multi-agent commitment-based protocols.

Montoya and Ovalle (2013) applied multi-agent systems for energy consumption by positioning a reactive inside wireless sensor networks, while Khalilian (2013) applied multi agent systems and data mining approaches towards a smart advisor's framework. Liu et al. (2012b) applied, successfully, multi-agent systems to bidding mechanism in an electricity auction.

In this chapter, the agent architecture implemented is shown in Fig. 11.1 and the multi-agent system is shown in Fig. 11.2. It has intelligence capability, which is a committee of a combination of multi-layer perceptrons and radial basis function network.

The agent is able to adapt using genetic programming, by adapting the committee structure. The next section describes neural networks which are applied to enable the agent to be intelligent.

Fig. 11.1 Illustration
of an agent

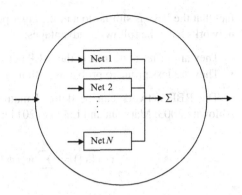

Fig. 11.2 Illustration
of a multi-agent system

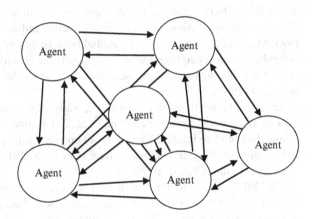

11.4 Neural Networks

This section describes neural networks which are used to model data. Neural
networks are, by definition, mathematical models that are inspired by the way
the human brain processes information. This section describes the type of neural
networks that relate some information to another, and these are called supervised
neural networks. Supervised neural networks take input data x and relate this to the
output data y. In this chapter, we apply two types of supervised neural networks and
these are radial basis functions and multi-layer pereceptron.

Radial basis function (RBF) is a neural network technique which is based on the
distance of the data set from its origin (Bishop 1995). The RBF is usually structured
with a single hidden layer of units with an activation function that is chosen from
a type of functions called basis functions. The activation of the hidden units is
characterized by a non-linear function of the distance between the input vector and
a vector indicating the centers of gravity of the data (Bishop 1995). Despite the

fact that the RBF is similar to a multi-layer perceptron (MLP), radial basis function networks have the following advantages:

- They are faster to train than the MLP networks
- They are less prone to problems with non-stationary inputs

The RBF network can be defined mathematically as follows (Buhmann and Ablowitz 2003; Marwala and Lagazio 2011):

$$y_k\left(\{x\}\right) = \sum_{j=1}^{M} w_{jk}\phi\left(\left\|\{x\} - \{c\}_j\right\|\right) \tag{11.1}$$

where, w_{jk} are the output weights, relating a hidden unit and an output unit, M shows the number of hidden units, $\{c\}_j$ is the center for the jth neuron, $\phi\left(\{x\}\right)$ is the jth non-linear activation function, $\{x\}$ is the input vector, and $k = 1, 2, 3, \ldots, M$ (Bishop 1995; Marwala and Lagazio 2011). Radial basis functions are trained in this chapter using the k-nearest neighbor method to estimate the centers and the weights are then estimated using the pseudo-inverse technique, and the details of these can be found in Bishop (1995).

Radial basis functions have been successfully applied to many complex problems such as voice transformation (Nirmal et al. 2013), image analysis of deformation (Biancolini and Salvini 2012), analysis of hemodynamics pattern flow (Ponzini et al. 2012), analysis of gene expression data (Liu et al. 2012a), and the prediction of logistics demand (Chen et al. 2012).

The MLP is a feed-forward neural network technique that approximates a relationship between sets of input data and a set of output data. It applies three or more layers of neurons, also called nodes, with non-linear activation functions. It can distinguish data that is not linearly separable or separable by a hyper-plane.

The MLP neural network consists of multiple layers of computational components normally inter-connected in a feed-forward manner (Haykin 1999; Hassoun 1995; Marwala 2012). Every neuron in one layer is connected to the neurons of the subsequent layer and this can be mathematically represented as follows (Haykin 1999):

$$y_k = f_{outer}\left(\sum_{j=1}^{M} w_{kj}^{(2)} f_{inner}\left(\sum_{i=1}^{d} w_{ji}^{(1)} x_i + w_{j0}^{(1)}\right) + w_{k0}^{(2)}\right) \tag{11.2}$$

Here, $w_{ji}^{(1)}$ and $w_{ji}^{(2)}$ are weights in the first and second layers, correspondingly, from input i to hidden unit j, M is the number of hidden units, d is the number of output units, while $w_{j0}^{(1)}$ and $w_{k0}^{(2)}$ are the weight parameters that indicate the biases for the hidden unit j and the output unit k. These weight parameters can be viewed as a mechanism that enables the model to understand the data. The weight vector in Eq. 11.2 is identified using the scaled conjugate gradient technique that is based on the maximum-likelihood method (Møller 1993).

The MLP has been successfully applied in many areas and these include power transformer diagnosis (Souahlia et al. 2012), automatic musical intrument recognition (Azarloo and Farokhi 2012), diagnosing of cervical cancer (Sokouti et al. 2012), automatic vehilce type classification (Daya et al. 2012), fingerprint spoof detection (Pereira et al. 2012), and intrusion detection (Ahmad et al. 2011).

The agent proposed in Fig. 11.1 contains a group of neural networks that collectively make a decision, and this is either called a committee approach or an ensemble of networks and is the subject of the next section.

11.5 Ensembles of Networks

When a group of neural networks are used to collectively make a decision, then this is known as an ensemble approach. There are many types of ensembles and, in this chapter, we discuss few of these and these are: bagging, boosting, stacking, and evolutionary committees.

11.5.1 Bagging

Bagging is a method that is based on an amalgamation of models fitted to bootstrap samples of a training data set to decrease the variance of the prediction model (Breiman 1996). Bagging fundamentally involves randomly selecting a section of the training data, training a model with this selection, and then iterating this procedure and then all trained models are pooled together with equal weights to form an ensemble. Bagging has been successfully applied in many areas such as the detection of obsessive compulsive disorder (Parrado-Hernández et al. 2012), diagnosing of arrhythmia beats (Mert et al. 2012), fraud detection tools (Louzada and Ara 2012), identification of MicroRNA Precursors (Jha et al. 2012), land-cover classification (Ghimire et al. 2012), and intrusion detection (Syarif et al. 2012).

11.5.2 Boosting

Boosting is a method that incrementally constructs an ensemble by training each new model with data that the heretofore trained model misclassified. Then the ensemble, which is a combination of all trained models, is used for prediction. Jasra and Holmes (2011) successfully applied stochastic boosting algorithms which used sequential Monte Carlo methods, while Leitenstorfer and Tutz (2011) successfully applied boosting methods to estimate single-index models. Other successful applications of boosting include object classification (Piro et al. 2013), categorization of natural scenes (Nock et al. 2012), automatic anatomy detection (Tajbakhsh et al. 2012), multi-view face pose classification (Yun and Gu 2012), and automatic audio tagging (Foucard et al. 2012).

11.5.3   Stacking

The general approach in mathematical modeling is that one chooses from a set of models by comparing them on data that was not used to train the models. This insight can also be applied to choose a model using a method called cross-validation (Bishop 1995). This is achieved by apportioning the data set into a *held-in* data set, which is used to train the models, and a *held-out* data set which is used to test the trained models (Sill et al. 2009; Marwala 2012).

Stacking uses performance of the model on the held-out data to combine the models instead of selecting from them the best performing model when tested on the held-out data and this gives an ensemble that performs better than any single one of the trained models (Wolpert 1992). Stacking has been successfully applied to many areas such as instance-based ensemble learning algorithms (Homayouni et al. 2010), real estate appraisal (Graczyk et al. 2010), and metabonomic applications (Lienemann et al. 2009).

11.5.4   Evolutionary Committees

Evolutionary committees are methods that are adaptive techniques that adapt to the environmental changes. This is usually achieved by evolving the weighting function that defines the contribution of each individual technique, with respect to the overall outcome of the committee.

Marwala (2009) introduced committees of networks for missing data estimation. The first committee of networks was made of multi-layer perceptrons (MLPs), support vector machines (SVMs), and radial basis functions (RBFs); and entailed the weighted combination of these three networks. The second, third, and fourth committees of networks were evolved using a genetic programming method and used the MLPs, RBFs and SVMs, respectively. The committees of networks were applied, collectively, with a hybrid particle swarm optimization and genetic algorithm technique for missing data estimation. When they were tested on an artificial taster, as well as HIV datasets, and then compared to the individual MLPs, RBFs, and SVMs for missing data estimation, the committee of networks approach was observed to give better results than the three approaches acting in isolation. Nonetheless, this improvement came at a higher computational load than the individual methods. In addition, it was observed that evolving a committee technique was a good way of constructing a committee.

In this chapter, we apply the three member ensemble which is shown in Fig. 11.3. The ideas presented in this section are an adaptation of the work done by Perrone and Cooper (1993) where they introduced the concept of a committee of networks and confirmed that this committee provides results that are more reliable than when using networks in isolation.

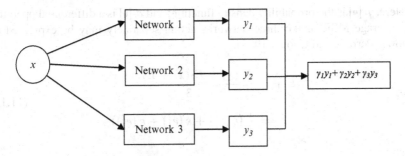

Fig. 11.3 Illustration of committee of networks

The mapping of the input x and output y can be expressed as the desired function plus an error. For notational accessibility, the mapping functions are assumed to have single outputs y_1, y_2, and y_3. This can be easily adapted to multiple outputs as follows (Perrone and Cooper 1993):

$$y_1(x) = h(x) + e_1(x) \tag{11.3}$$

$$y_2(x) = h(x) + e_2(x) \tag{11.4}$$

$$y_3(x) = h(x) + e_3(x) \tag{11.5}$$

Here, $h(\cdot)$ is the estimated mapping function; and $e(\cdot)$ is the error.

The mean square errors (MSE) for model $y_1(x)$, $y_2(x)$, and $y_3(x)$ may be expressed as follows (Perrone and Cooper 1993):

$$E_1 = \varepsilon\left[\{y_1(x) - h(x)\}^2\right] = \varepsilon\left[e_1^2\right] \tag{11.6}$$

$$E_2 = \varepsilon\left[\{y_2(x) - h(x)\}^2\right] = \varepsilon\left[e_2^2\right] \tag{11.7}$$

$$E_3 = \varepsilon\left[\{y_2(x) - h(x)\}^2\right] = \varepsilon\left[e_3^2\right] \tag{11.8}$$

Here, $\varepsilon[\bullet]$ denotes the expected value and corresponds to the integration over the input data, and is defined as follows (Perrone and Cooper 1993):

$$\varepsilon\left[e_1^2\right] \equiv \int e_1^2(x)p(x)dx \tag{11.9}$$

$$\varepsilon\left[e_2^2\right] \equiv \int e_2^2(x)p(x)dx \tag{11.10}$$

$$\varepsilon\left[e_3^2\right] \equiv \int e_3^2(x)p(x)dx \tag{11.11}$$

Here, $p\, [\bullet]$ is the probability density function; and $d\, [\bullet]$ is a differential operator. The average MSE of the three networks acting separately may be expressed as follows (Perrone and Cooper 1993):

$$E_{AV} = \frac{E_1(x) + E_2(x) + E_3(x)}{3}$$

$$= \frac{1}{3} \left(\varepsilon \left(e_1^2 \right) + \varepsilon \left(e_2^2 \right) + \varepsilon \left(e_3^2 \right) \right) \tag{11.12}$$

11.5.4.1 Equal Weights

The output of the committee is the average of the outputs from the three networks. The committee prediction may be expressed in the following form, by giving equal weighting functions (Perrone and Cooper 1993):

$$y_{COM} = \frac{1}{3} \left(y_1(x) + y_2(x) + y_3(x) \right) \tag{11.13}$$

The MSE of the committee can be written as follows:

$$E_{COM} = \varepsilon \left[\left(\frac{1}{3} \{ y_1(x) + y_2(x) + y_3(x) \} - \frac{1}{3} [h(x) + h(x) + h(x)] \right)^2 \right]$$

$$= \varepsilon \left[\left(\frac{1}{3} \{ [y_1(x) - h(x)] + [y_2(x) - h(x)] + [y_3(x) - h(x)] \} \right)^2 \right]$$

$$= \varepsilon \left[\left(\frac{1}{3} \{ e_1 + e_2 + e_3 \} \right)^2 \right]$$

$$= \frac{1}{9} \left(\varepsilon \left[e_1^2 \right] + 2 \left(\varepsilon \left[e_1 e_2 \right] + \varepsilon \left[e_1 e_2 \right] + \varepsilon \left[e_2 e_3 \right] + \varepsilon \left[e_1 e_3 \right] \right) + \varepsilon \left[e_2^2 \right] + \varepsilon \left[e_3^2 \right] \right) \tag{11.14}$$

If it is assumed that the errors $(e_1, e_2,$ and $e_3)$ are uncorrelated then

$$\varepsilon[e_1 e_2] = \varepsilon[e_1 e_2] = \varepsilon[e_2 e_3] = \varepsilon[e_1 e_3] = 0 \tag{11.15}$$

Substituting Eq. 11.15 in Eq. 11.14, the error of the committee can be related to the average error of the networks acting individually as follows (Perrone and Cooper 1993):

$$E_{COM} = \frac{1}{9} \left(\varepsilon \left[e_1^2 \right] + \varepsilon \left[e_2^2 \right] + \varepsilon \left[e_3^2 \right] \right)$$

$$= \frac{1}{3} E_{AV} \tag{11.16}$$

Equation 11.16 indicates that the MSE of the committee is one-third of the average MSE of the individual technique. This implies that the MSE of the committee is always equal to or less than the average MSE of the three methods acting individually.

11.5.4.2 Variable Weights

The three networks might not essentially have the same predictive capability. To accommodate the strength of each technique, the network should be given suitable weighting functions. It will be explained later how these weighting functions will be evaluated when there is no prior knowledge of the strength of each approach.

The output of the ensemble may be defined as the combination of the three independent methods with estimated weighting functions as:

$$y_{COM} = \gamma_1 y_1(x) + \gamma_2 y_2(x) + \gamma_3 y_3(x) \tag{11.17}$$

where γ_1, γ_2, and γ_3 are the weighting functions and $\gamma_1 + \gamma_2 + \gamma_3 = 1$. The MSE due to the weighted committee can be written as follows (Marwala 2000):

$$
\begin{aligned}
E_{COM} &= \varepsilon \left[(\gamma_1 y_1(x) + \gamma_2 y_2(x) + \gamma_3 y_3(x) - [\gamma_1 h(x) + \gamma_2 h(x) + \gamma_3 h(x)])^2 \right] \\
&= \varepsilon \left[(\gamma_1 [y_1(x) - h(x)] + \gamma_2 [y_2(x) - h(x)] + \gamma_3 [y_3(x) - h(x)])^2 \right] \\
&= \varepsilon \left[(\gamma_1 e_1 + \gamma_2 e_2 + \gamma_3 e_3)^2 \right]
\end{aligned}
\tag{11.18}
$$

Equation 11.18 may be rewritten in Lagrangian form as follows (Perrone and Cooper 1993):

$$E_{COM} = \varepsilon \left[(\gamma_1 e_1 + \gamma_2 e_2 + \gamma_3 e_3)^2 \right] + \lambda (1 - \gamma_1 - \gamma_2 - \gamma_3) \tag{11.19}$$

where λ is the Lagrangian multiplier. The derivative of the error in Eq. 11.19 with respect to γ_1, γ_2, γ_3 and λ may be calculated and equated to zero as follows (Perrone and Cooper 1993):

$$\frac{dE_{COM}}{d\gamma_1} = 2e_1 \varepsilon [(\gamma_1 e_1 + \gamma_2 e_2 + \gamma_3 e_3)] - \lambda = 0 \tag{11.20}$$

$$\frac{dE_{COM}}{d\gamma_2} = 2e_2 \varepsilon [(\gamma_1 e_1 + \gamma_2 e_2 + \gamma_3 e_3)] - \lambda = 0 \tag{11.21}$$

$$\frac{dE_{COM}}{d\gamma_3} = 2e_3 \varepsilon [(\gamma_1 e_1 + \gamma_2 e_2 + \gamma_3 e_3)] - \lambda = 0 \tag{11.22}$$

$$\frac{d\,E_{COM}}{d\lambda} = 1 - \gamma_1 - \gamma_2 - \gamma_3 = 0 \qquad (11.23)$$

In solving Eqs. 11.20, 11.21, 11.22, and 11.23, the minimum error is obtained when the weights are (Perrone and Cooper 1993):

$$\gamma_1 = \frac{1}{1 + \dfrac{\varepsilon\left[e_1^2\right]}{\varepsilon\left[e_2^2\right]} + \dfrac{\varepsilon\left[e_1^2\right]}{\varepsilon\left[e_3^2\right]}} \qquad (11.24)$$

$$\gamma_2 = \frac{1}{1 + \dfrac{\varepsilon\left[e_2^2\right]}{\varepsilon\left[e_1^2\right]} + \dfrac{\varepsilon\left[e_2^2\right]}{\varepsilon\left[e_3^2\right]}} \qquad (11.25)$$

$$\gamma_3 = \frac{1}{1 + \dfrac{\varepsilon\left[e_3^2\right]}{\varepsilon\left[e_1^2\right]} + \dfrac{\varepsilon\left[e_3^2\right]}{\varepsilon\left[e_2^2\right]}} \qquad (11.26)$$

Equations 11.24, 11.25, and 11.26 may be generalized for a committee with n-trained networks and may be written as follows (Perrone and Cooper 1993):

$$\gamma_i = \frac{1}{\displaystyle\sum_{j=1}^{n} \dfrac{\varepsilon\left[e_i^2\right]}{\varepsilon\left[e_j^2\right]}} \qquad (11.27)$$

From Eq. 11.27, the following conditions may be derived as follows (Marwala 2000):

$$\varepsilon\left[e_1^2\right] = \varepsilon\left[e_2^2\right] = \varepsilon\left[e_3^2\right] \Rightarrow \gamma_1 = \gamma_2 = \gamma_3 = \frac{1}{3} \qquad (11.28)$$

$$\varepsilon\left[e_3^2\right] < \varepsilon\left[e_2^2\right] < \varepsilon\left[e_1^2\right] \Rightarrow \gamma_1 < \gamma_2 < \gamma_3; \gamma_3 > \frac{1}{3} \qquad (11.29)$$

$$\varepsilon\left[e_1^2\right] < \varepsilon\left[e_2^2\right] < \varepsilon\left[e_3^2\right] \Rightarrow \gamma_3 < \gamma_2 < \gamma_3; \gamma_1 > \frac{1}{3} \qquad (11.30)$$

11.6 Genetic Algorithms

The multi-agent system proposed in this chapter is adaptive and this is enabled by a genetic algorithm. Genetic algorithms were enthused by Darwin's theory of natural evolution. In natural evolution, members of a population compete with each other to survive and reproduce. Evolutionary successful individuals reproduce, while weaker members disappear. Consequently, the genes that are successful are probably going to spread within the population. This natural optimization technique is applied in this chapter to optimize the decision of a committee of agents shown in Fig. 11.3. This essentially allows an agent to become better, based on how well it performed and, in this chapter, on trading in the stock market.

The basic genetic algorithm proposed by Holland (1975) is applied. The algorithm acts on a population of binary-string chromosomes. These chromosomes are acquired by utilizing the Gray algorithm. Each of these strings is a discretized representation of a point in the search space. Here we are searching for the most optimum combination of architectures that form a committee and that give the least errors. Consequently, the fitness function is the error offered by committee of agents. On producing a new population, three operators are executed: (1) crossover; (2) mutation; (3) and reproduction.

Similar to natural evolution, the probability of mutation happening is lower than that of crossover or reproduction. The crossover operator combines genetic information in the population by cutting pairs of chromosomes at random points along their length and exchanging over the cut sections. This operator has a potential of connecting successful operators together. Simple crossover is applied in this chapter. The mutation operator picks a binary digit of the chromosomes at random and inverts it. This has the potential of introducing to the population new information. Reproduction takes successful chromosomes and reproduces them in accordance to their fitness function. The fit parameters are allowed to reproduce and the weaker parameters are removed. This is conducted using the roulette wheel procedure.

Genetic algorithms have been applied successfully in many areas such as content based image retrieval (Syam and Rao 2013), variable selection in solar radiation estimation (Will et al. 2013), non-destructive characterization of tie-rods (Gentilini et al. 2013), assembly line worker assignment (Mutlu et al. 2013), sheep farming (Del Pilar Angulo-Fernandez et al. 2013), and power consumption (Shen and Zhang 2013).

11.7   Simulating the Stock Marketing

In this chapter, we apply a multi-agent system to model the stock market. Multi-agent systems have been applied to stock markets in the past (Tirea et al. 2012; Liu and Cao 2011; Yoshikazu and Shozo 2007; Ikeda and Tokinaga 2004). The structure that is proposed consists of committees of agents forming a player in the stock market. The simulation framework consists of a population of these players that compete for a fixed number of stocks. The agents learn through the use of neural networks. The structure of each agent evolves using a genetic algorithm such that its contribution to the overall function of a committee adapts to the evolutionary time-varying nature of the problem. The characteristics of the agents that evolve are the number of hidden units and the weight contribution of each network towards a player. The number of hidden units is constrained to fall within a given space, in this study 1 and 10. Each committee of agents, known as a player, trades stocks with other agents and when prices of stocks are announced, the players trade by following these rules:

- Once a price is announced, the committees look at the current price and the future price of stocks. The future price is determined from the agents that learn using neural networks. For a player, the predicted price is the average of the prediction of each agent within that particular player.
- If the predicted price of a stock is lower than the current price, then the player tries to sell the stock. If the predicted price for the stock is higher than the current price, then the committee tries to buy the stock.
- At any given stage, the committee is only prepared to sell the maximum of 40 % of the volume of stocks it has.
- The amount of stocks that a committee buys or sells depends on, amongst other factors, the predicted price. If the predicted price of a particular stock is x % higher than the current price, the committee tries to acquire x % of the volume available on the market of that particular stock. This simulation is started by choosing the number of players that participate in the trading of stocks, together with the number of agents that form a player. Then the agents are trained by randomly assigning the number of hidden units to fall in the interval [1 10] and assigning weighting functions of the committee. The agents are trained using the data from the previous 50 trading days. The trained agents are grouped into their respective players and are then used to predict the next price, given the current price. The simulation followed in this chapter is shown in Fig. 11.4.

After 50 days of trading have elapsed, the performance of each agent and the weighting functions are evaluated and these are transformed into 8 bits and each player exchanges bits with other players, a process called crossover. Thereafter, the agents mutate at low probability. The successful agents are duplicated, while the less successful ones are eliminated. Then the networks are retrained again and the whole process is repeated. When a price is announced, trading of stocks is conducted until the consensus is reached. At this state, the overall wealth of the committees does not increase as a result of trading.

The example that is considered in this study is the trading of three stocks, namely: (1) the Dow Jones; (2) NASDAQ; and (3) S&P 500. The time-histories of the stocks are downloaded from the Internet and used to train agents. For a given set of price of these stocks, the committee of agents predicts the future prices of stocks. It should be noted that, on implementing this procedure, the total number of stocks available is kept constant. This figure indicates that sometimes the players with successful strategies do not necessarily dominate indefinitely. This is due to the fact that, strategies that are successful in one time frame are not necessarily successful at a later time.

When the scalability of the simulations was studied, it was found that the method proposed was scalable. However, it was observed that the computational time increased with the increase in the number of agents and players. A linear relationship existed between the average computational time taken to run the complete simulation and the number of players as well as the number of agents that form a player.

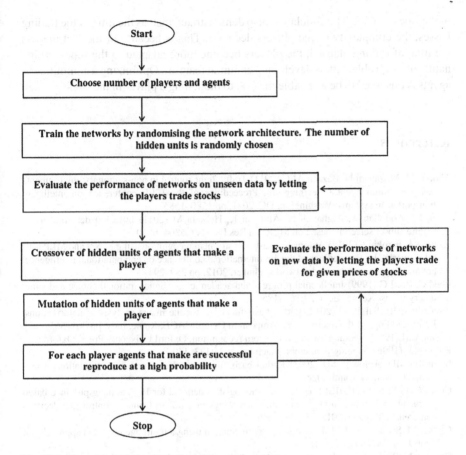

Fig. 11.4 Illustration of the simulation of the stock market

The complexity of the populations of agents that make players of the game was studied and defined as the measure of a degree of variation in a population of agents. Each species of agents form a dimension in space. Each dimension has a variation indicating the level of complexity of a population of that species. The results indicated that, as the system evolved, the number of hidden units for a given player steadily decreased and stabilized around 3. It was additionally observed that no player had the monopolistic advantage on the prediction of the stock market.

11.8   Conclusions

A simulation of the stock market was successfully implemented. It is established that the number of players and agents that form a player that partake in the trading game are directly proportional to the computational time taken to run the simulation. It is additionally found that, no player has the monopolistic advantage on the prediction of the stock market. The simulation also demonstrated that, as the time of the trading passes, the complexity of the players decrease. This is because of the fact that, as the time of trading elapsed, the players become more adapted to the time-varying nature of the problem, thus developing common features. Optimizing a committee of agents is observed to be a feasible method to modelling a player in the stock market.

References

Abdoos M, Mozayani N, Bazzan ALC (2011) Traffic light control in non-stationary environments based on multi agent Q-learning. In: Proceedings of the IEEE conference on intelligent transportation systems, Washington, DC, 2011, pp 1580–1585

Ahmad I, Abdullah A, Alghamdi A, Alnfajan K, Hussain M (2011) Intrusion detection using feature subset selection based on MLP. Sci Res Essays 6:6804–6810

Azarloo A, Farokhi F (2012) Automatic musical instrument recognition using K-NN and MLP neural networks. In: Proceedings of the 4th international conference on computational intelligence, communication systems and networks, Phuket, 2012, pp 289–294

Beed C, Beed C (1999) Intellectual progress and academic economics: rational choice and game theory. J Post Keynes Econ 22:163–185

Biancolini ME, Salvini P (2012) Radial basis functions for the image analysis of deformations. In: Proceedings of the international symposium CompIMAGE, Rome, Italy, pp 361–365

Bishop CM (1995) Neural networks for pattern recognition. Oxford University Press, Oxford

Breiman L (1996) Bagging predictors. Mach Learn 24:123–140

Buhmann MD, Ablowitz MJ (2003) Radial basis functions: theory and implementations. Cambridge University, Cambridge

Chen Y, Hu S, Liu H (2012) Prediction of the logistics demand for Heilongjiang province based on radial basis function algorithm. In: Proceedings of the 24th Chinese control and decision conference, Taiyuan, 2012, pp 2358–2361

Chitsaz M, Seng W (2013) Medical image segmentation using a multi-agent system approach. Int Arab J Inf Technol 10 (in press)

Christin N (2011) Network security games: combining game theory, behavioral economics and network measurements. Lect Notes Comput Sci 7037:4–6

Daya B, Akoum AH, Bahlak S (2012) Geometrical features for multiclass vehicle type recognition using MLP network. J Theor Appl Inf Technol 43:285–294

Del Pilar A-FI, Aguilar-Lasserre AA, Gonzalez-Huerta MA, Moras-Sanchez CG (2013) Investing in the sheep farming industry: a study case based on genetic algorithms. Lect Notes Electr Eng 130:337–346

Elammari M, Issa Z (2013) Using model driven architecture to develop multi-agent systems. Int Arab J Inf Technol 10 (in press)

El-Menshawy M, Bentahar J, El Kholy W, Dssouli R (2013) Verifying conformance of multi-agent commitment-based protocols. Expert Syst Appl 40:122–138

Foucard R, Essid S, Lagrange M, Richard G (2012) A regressive boosting approach to automatic audio tagging based on soft annotator fusion. In: Proceedings of the IEEE international conference on acoustics, speech and signal processing, Kyoto, 2012, pp 73–76

Franklin S, Graesser A (1996) Is it an agent, or just a program?: A taxonomy for autonomous agents. In: Proceedings of the third international workshop on agent theories, architectures, and languages, pp 21–35

Gentilini C, Marzani A, Mazzotti M (2013) Nondestructive characterization of tie-rods by means of dynamic testing, added masses and genetic algorithms. J Sound Vib 332:76–101

Ghimire B, Rogan J, Galiano V, Panday P, Neeti N (2012) An evaluation of bagging, boosting, and random forests for land-cover classification in Cape Cod, Massachusetts, USA. GIScience Remote Sens 49:623–643

Goldberg DE (1989) Genetic algorithms in search, optimization and machine learning. Addison-Wesley, Reading

Graczyk M, Lasota T, Trawiński B, Trawiński K (2010) Comparison of bagging, boosting and stacking ensembles applied to real estate appraisal. Lect Notes Comput Sci 5991:340–350

Hanauske M, Kunz J, Bernius S, König W (2010) Doves and hawks in economics revisited: an evolutionary quantum game theory based analysis of financial crises. Physica A Stat Mech Appl 389:5084–5102

Hassoun MH (1995) Fundamentals of artificial neural networks. MIT Press, Cambridge, MA

Haykin S (1999) Neural networks. Prentice-Hall, Upper Saddle River

Hodgson GM, Huang K (2012) Evolutionary game theory and evolutionary economics: are they different species? J Evol Econ 22:345–366

Holland J (1975) Adaptation in natural and artificial systems. University of Michigan Press, Ann Arbor

Homayouni H, Hashemi S, Hamzeh A (2010) Instance-based ensemble learning algorithm with stacking framework. In: Proceedings of the 2nd international conference on software technology and engineering, San Juan, 2010, pp 164–169

Hui ECM, Bao H (2013) The logic behind conflicts in land acquisitions in contemporary China: a framework based upon game theory. Land Use Policy 30:373–380

Hurwitz E, Marwala T (2007) Learning to bluff. http://arxiv.org/ftp/arxiv/papers/0705/0705.0693.pdf. Last Accessed 7 Mar 2013

Ikeda Y, Tokinaga S (2004) Chaoticity and fractality analysis of an artificial stock market generated by the multi-agent systems based on the co-evolutionary genetic programming. IEICE Trans Fundam Electron Commun Comput Sci 87(9):2387–2394

Jasra A, Holmes CC (2011) Stochastic boosting algorithms. Stat Comput 21:335–347

Jha A, Chauhan R, Mehra M, Singh HR, Shankar R (2012) miR-BAG: bagging based identification of microRNA precursors. PLoS One 7:e45782

Kasabov N (1998) Introduction: hybrid intelligent adaptive systems. Int J Intell Syst 6:453–454

Khalilian M (2013) Towards smart advisor's framework based on multi agent systems and data mining methods. Lect Notes Electr Eng 156:73–78

Laffont J-J (1997) Game theory and empirical economics: the case of auction data. Eur Econ Rev 41:1–35

Leitenstorfer F, Tutz G (2011) Estimation of single-index models based on boosting techniques. Stat Model 11:203–217

Lienemann K, Plötz T, Fink GA (2009) Stacking for ensembles of local experts in metabonomic applications. Lect Notes Comput Sci 5519:498–508

Liu X, Cao H (2011) Price limit and the stability of stock market: an application based on multi-agent system. In: Proceedings of the 2nd international conference on artificial intelligence management science and electronic commerce, Deng Leng, 2011, pp 484–487

Liu W, Ji Z, He S, Zhu Z (2012a) Survival analysis of gene expression data using PSO based radial basis function networks. In: Proceedings of the IEEE congress on evolutionary computation, Brisbane, 2012, art. no. 6256144, pp 1–5

Liu Z, Yan J, Shi Y, Zhu K, Pu G (2012b) Multi-agent based experimental analysis on bidding mechanism in electricity auction markets. Int J Electr Power Energy Syst 43:696–702

Louzada F, Ara A (2012) Bagging k-dependence probabilistic networks: an alternative powerful fraud detection tool. Expert Syst Appl 39:11583–11592

Marwala T (2000) On damage identification using a committee of neural networks. J Eng Mech 126:43–50

Marwala T (2009) Computational intelligence for missing data imputation, estimation and management: knowledge optimization techniques. IGI Global Publications, New York

Marwala T (2010) Finite element model updating using computational intelligence techniques. Springer, London

Marwala T (2012) Condition monitoring using computational intelligence methods. Springer, London

Marwala T, Lagazio M (2011) Militarized conflict modeling using computational intelligence techniques. Springer, London

Marwala T, De Wilde P, Correia L, Mariano P, Ribeiro R, Abramov V, Szirbik N, Goossenaerts J (2001) Scalability and optimisation of a committee of agents using genetic algorithm. In: Proceedings of the 2001 international symposia on soft computing and intelligent systems for industry, Parsley, Scotland, arxiv 0705.1757

McCain RA (2009) Commitment and weakness of will in game theory and neoclassical economics. J Socio-Econ 38:549–556

McFadden DW, Tsai M, Kadry B, Souba WW (2012) Game theory: applications for surgeons and the operating room environment. Surgery (United States) 152:915–922

Mert A, Kiliç N, Akan A (2012) Evaluation of bagging ensemble method with time-domain feature extraction for diagnosing of arrhythmia beats. Neural Comput Appl:1–10. doi:10.1007/s00521-012-1232-7

Michalewicz Z (1996) Genetic algorithms + data structures = evolution programs. Springer, London

Møller AF (1993) A scaled conjugate gradient algorithm for fast supervised learning. Neural Netw 6:525–533

Montoya A, Ovalle D (2013) Energy consumption by deploying a reactive multi-agent system inside wireless sensor networks. Lect Notes Electr Eng 152:925–934

Mutlu Ö, Polat O, Supciller AA (2013) An iterative genetic algorithm for the assembly line worker assignment and balancing problem of type-II. Comput Oper Res 40:418–426

Nirmal JH, Patnaik S, Zaveri MA (2013) Voice transformation using radial basis function. Lect Notes Electr Eng 150:345–351

Nock R, Piro P, Nielsen F, Bel Haj Ali W, Barlaud M (2012) Boosting k-NN for categorization of natural scenes. Int J Comput Vis 100:294–314

Papadimitriou CH (2001) Game theory and mathematical economics: a theoretical computer scientist's introduction. In: Proceedings of the annual symposium on foundations of computer science, Las Vegas, Nevada, pp 4–8

Parrado-Hernández E, Gómez-Verdejo V, Martínez-Ramón M, Shawe-Taylor J, Alonso P, Pujol J, Menchón JM, Cardoner N, Soriano-Mas C (2012) Voxel selection in MRI through bagging and conformal analysis: application to detection of obsessive compulsive disorder. In: Proceedings of the 2nd international workshop on pattern recognition in neuroimaging, London, 2012, pp 49–52

Pereira LFA, Pinheiro HNB, Silva JIS, Silva AG, Pina TML, Cavalcanti GDC, Ren TI, De Oliveira JPN (2012) A fingerprint spoof detection based on MLP and SVM. In: Proceedings of the international joint conference on neural networks, Brisbane, 2012, art. no. 6252582, pp 1–7

Perrone MP, Cooper LN (1993) When networks disagree: ensemble methods for hybrid neural networks. In: Mammone RJ (ed) Artificial neural networks for speech and vision. Chapman and Hall, London

Piro P, Barlaud M, Nock R, Nielsen F (2013) K-NN boosting prototype learning for object classification. Lect Notes Electr Eng 158:37–53

Ponzini R, Biancolini ME, Rizzo G, Morbiducci U (2012) Radial basis functions for the interpolation of hemodynamics flow pattern: a quantitative analysis. In: Proceedings of the international symposium CompIMAGE, Rome, Italy, pp 341–345

Ross D (2006) Evolutionary game theory and the normative theory of institutional design: binmore and behavioral economics. Polit Philos Econ 5:51–79

Roth AE (2002) The economist as engineer: game theory, experimentation, and computation as tools for design economics. Econometrica 70:1341–1378

Roy K, Kamel MS (2012) Facial expression recognition using game theory. Lect Notes Comput Sci 7477:139–150

Russell SJ, Norvig P (2003) Artificial intelligence: a modern approach. Prentice Hall, Upper Saddle River

Shen G, Zhang Y (2013) Power consumption constrained task scheduling using enhanced genetic algorithms. Stud Comput Intell 432:139–159

Sill J, Takacs G, Mackey L, Lin D (2009) Feature-weighted linear stacking. http://arxiv.org/abs/0911.0460. Last Accessed 7 Mar 2013

Sokouti B, Haghipour S, Tabrizi AD (2012) A framework for diagnosing cervical cancer disease based on feedforward MLP neural network and ThinPrep histopathological cell image features. Neural Comput Appl:1–12. doi:10.1007/s00521-012-1220-y

Souahlia S, Bacha K, Chaari A (2012) MLP neural network-based decision for power transformers fault diagnosis using an improved combination of Rogers and Doernenburg ratios DGA. Int J Electr Power Energy Syst 43:1346–1353

Stroeve SH, Blom HAP, Bakker GJ (2013) Contrasting safety assessments of a runway incursion scenario: event sequence analysis versus multi-agent dynamic risk modelling. Reliab Eng Syst Saf 109:133–149

Syam B, Rao YS (2013) An effective similarity measure via genetic algorithm for content based image retrieval with extensive features. Int Arab J Inf Technol 10 (in press)

Syarif I, Zaluska E, Prugel-Bennett A, Wills G (2012) Application of bagging, boosting and stacking to intrusion detection. Lect Notes Comput Sci 7376:593–602

Tajbakhsh N, Wu H, Xue W, Gotway MB, Liang J (2012) A novel online boosting algorithm for automatic anatomy detection. Mach Vis Appl: (26 October 2012), pp 1–12. doi:10.1007/s00138-012-0455-z

Teweldemedhin E, Marwala T, Mueller C (2004) Agent-based modelling: a case study in HIV epidemic. In: Proceedings of the fourth international conference on hybrid intelligent systems, Tokyo, Japan, pp 154–159

Tirea M, Tandau I, Negru V (2012) Stock market multi-agent recommendation system based on the Elliott Wave Principle. Lect Notes Comput Sci 7465:332–346

van den Brink R, van der Laan G, Vasil'ev V (2008) Extreme points of two digraph polytopes: description and applications in economics and game theory. J Math Econ 44:1114–1125

Villena MG, Villena MJ (2004) Evolutionary game theory and Thorstein Veblen's evolutionary economics: is EGT Veblenian? J Econ Issues 38:585–610

Wei X, Qu H, Ma E (2012) Decisive mechanism of organizational citizenship behavior in the hotel industry – an application of economic game theory. Int J Hosp Manag 31:1244–1253

Will A, Bustos J, Bocco M, Gotay J, Lamelas C (2013) On the use of niching genetic algorithms for variable selection in solar radiation estimation. Renew Energy 50:168–176

Wolpert DH (1992) Stacked generalization. Neural Netw 5:241–259

Yoshikazu I, Shozo T (2007) Multi-fractality analysis of time series in artificial stock market generated by multi-agent systems based on the genetic programming and its applications. IEICE Trans Fundam Electron Commun Comput Sci 90(10):2212–2222

Yun Y, Gu IYH (2012) Multi-view face pose classification by boosting with weak hypothesis fusion using visual and infrared images. In: Proceedings of the IEEE international conference on acoustics, speech and signal processing, Kyoto, 2012, pp 1949–1952

Chapter 12
Control Approaches to Economic Modeling: Application to Inflation Targeting

Abstract In this chapter, a control system approach that is based on artificial intelligence is adopted to analyze the inflation targeting strategy. The input/output model is constructed using a multi-layered perceptron network and a closed loop control strategy is adopted using a genetic algorithm to control inflation through the manipulation of interest rates. Given the historical inflation rate data, a control scheme is used to determine the interest rate that is required to attain the given inflation rate. The calculated interest rate is then compared to the historical inflation rate to evaluate the effectiveness of the control strategy.

12.1 Introduction

Inflation is a measure of the rate of increase of the price levels. It is a major factor in many areas of economic life such as when negotiating with unions on salary increases. Many countries implement what is known as inflation targeting, which is essentially a policy framework that seeks to maintain inflation between certain bounds.

Countries' economies have been decimated by what is called hyper-inflation, which is a situation where the inflation levels are so high that the currency becomes useless. Recently, the economy of Zimbabwe experienced hyper-inflation which decimated its economy and ended up with the dollarization of its economy to replace its own currency.

There are many other similar experiences of hyper-inflation including in Brazil between 1967 and 1994, Germany in 1923 and this is thought to have ultimately resulted in the seizure of power by the Nazis and, subsequently, de-civilizing and plunging Germany into the dark ages, Hungary in 1945–1946, and many more other countries.

It is important to maintain some level of inflation in the economy because if the economy experiences deflation then rational consumers postpone the purchase of goods and services to a future date when the goods and services will be cheaper.

If, however, the economy experiences hyper-inflation then rational consumers seek to buy goods and services today because if they postpone the purchase, goods and services will be unaffordable in the future date. Both these scenarios are undesirable.

Inflation thus needs to be controlled. Amira et al. (2013) studied the growth effects of inflation targeting in the emerging markets and observed that, even though inflation targeting leads to higher economic growth, it does not automatically assure a stable growth rate.

Yilmazkuday (2013) studied inflation targeting and observed that during the Turkish inflation-targeting period regional inflation rates converged to each other with non-tradable constituents, while they diverged from each other with tradable constituents.

Carrasco and Ferreiro (2013) analyzed inflation expectations in Mexico and applied unit root, normality and co-integration tests. The results they obtained rejected the null hypothesis of normality for inflation expectations over the studied period. They demonstrated a persistent relationship between the exchange rate and the interest rate. They also observed that inflation expectations influence the long-term dynamics.

Karahan and Çolak (2013) studied the impact of the inflation targeting policy on the inflation uncertainty in Turkey. They applied the ARCH-GARCH technique to generate an inflation uncertainty series and observed that an inflation targeting policy was a great strategy for reducing inflation uncertainty.

Ding and Kim (2012) investigated whether inflation targeting matters for purchasing power parity (PPP). The results obtained demonstrated that inflation targeting contributes a vital contribution in giving favourable information for long-run PPP.

García-Solanes and Torrejón-Flores (2012) analyzed the macroeconomic effects of inflation targeting in five Latin American countries from 2000 to 2007. They observed that inflation gave to lower variability in gross domestic product (GDP) growth.

Bleich et al. (2012) approximated forward-looking monetary policy rules for 20 inflation targeting countries. They observed that inflation targeting considerably changes the central bank's response purpose of inflation stabilization. They also demonstrated that inflation targeting stabilizes inflation.

Pourroy (2012) studied whether the exchange rate control improved inflation targeting in emerging economies. They demonstrated considerable confirmation that inflation targeting is improved by the exchange rate.

Abo-Zaid and Tuzemen (2012) studied the effects of the inflation targeting on levels and volatilities of inflation, GDP growth and fiscal imbalances. They observed that inflation targeting in developing countries gave rise to lower and stable inflation as well as higher and stable GDP growth. In the developed nations, inflation targeting gave higher GDP growth.

Kose et al. (2012) studied the relationship between the interest rate and inflation rate in Turkey between 2002 and 2009. The results obtained demonstrated that monetary policy rates depended on inflationary expectations and that long-term interest rates depended on monetary policy.

Palma and Portugal (2009) studied the formation of inflationary expectations in Brazil using neural networks. The results obtained showed that inflationary expectations were most influenced by the exchange rate volatility and commodities prices variation.

Broto (2011) successfully applied the GARCH models to study inflation target-ing in Latin America, while Harjes and Ricci (2010) successfully applied Bayesian analysis for inflation targeting in South Africa. Atesoglu and Smithin (2006) contended that, an inflation-targeting policy was not an appropriate monetary policy rule. They argued that, inflation targeting reduced the equilibrium growth rate and that lower inflation target reduced real wages as well as profits and increased interest rates.

Lomax (2005) reviewed the function that model-based forecasts play in the monetary policy process in the United Kingdom. They observed that predictive models have been useful for inflation targeting.

Levin (2004) proposed a model of inflation targeting in a small open economy under floating exchange rates and applied Taylor rule to attain a target inflation rate and Phillips curve to determine the inflation process.

It is clear that the concept of inflation targeting is applicable and has been successful. In this chapter, we adopt a framework that involves modeling inflation and using the control framework for inflation targeting. To model inflation, we use a multi-layered perceptron neural network and, for control, we use a genetic algorithm. It is important to note that it is easier to model data that is stationary than to model data that is not stationary. Stationary data is a data set whose character does not change as a function of time. If the data is non-stationary, then it is difficult to model. The next section asks the question whether inflation data is stationary or not?

12.2   Is Inflation Non-stationary?

As explained in Chap. 2 of this book, a stationary process is a stochastic process whose joint probability distribution does not fluctuate when shifted in space or time. As a result, if certain properties of the data, such as the mean and variance, can be approximated then they do not change over space or time (Priestley 1988). A non-stationary process is a process whose joint probability distribution fluctuates when moved in space or time.

There are many techniques that have been proposed to evaluate whether a given signal is stationary or not and these include a quantification of similarities of the auto correlation integral of a subdivision of a time series and the cross-correlation of that sub-division with others of the same time series by Kiremire and Marwala (2008); Dickey-Fuller and Phillips-Perron Tests (Perron 1988), Kwiatkowski-Phillips-Schmidt-Shin Test (Kwiatkowski et al. 1992), as well as the Variance Ratio Test (Granger and Newbold 1974; Schwert 1989).

This section evaluates whether the inflation is stationary and applies the variance test to answer that question. As described in Chap. 2, the Variance Ratio Test quantifies the stationarity of a signal by calculating the variance ratio (F) as follows (Lo and MacKinlay 1989):

$$F = \frac{Ve}{Vu} \tag{12.1}$$

Here, V_e is the explained variance and V_u is the unexplained variance and:

$$V_e = \frac{\sum\limits_{i} n_i \left(\bar{X}_i - \bar{X} \right)^2}{K - 1} \tag{12.2}$$

and

$$V_u = \frac{\sum\limits_{ij} \left(X_{ij} - \bar{X}_i \right)^2}{N - K} \tag{12.3}$$

Here, $\bar{X}_i$ indicates the sample mean of the ith group, n_i is the number of observations in the ith group, a $\bar{X}$ indicates the complete mean of the data, X_{ij} is the jth observation in the ith out of K groups and N is the overall sample size. When the variability ratio is 1, then the data follows a random walk, if it is larger than 1, then it is non-stationary, and if it is less than 1, then it shows a mean reversal.

Russell (2011) contended that, for the reason that the United States inflation has been non-stationary over the past five decades, the enormous amount of empirical research that did not account for non-stationarity was unsound. They studied 50 years of United States inflation data and found that the Phillips curve results were as a result of the non-stationarity inflation over the period.

Lopez (2009) studied the stationary behavior of the inflation rates for the Euro-zone members and some bordering countries between 1957 and 2007 and observed that some of the Euro-zone inflation rates were non-stationary, while some were stationary.

Russell and Banerjee (2008) observed that theories of inflation include a Phillips curve and are generally approximated using methods that pay no attention to the non-stationary nature of inflation and, therefore, are inaccurate. They approximated a Phillips curve which factored non-stationarity in inflation and observed a positive relationship between inflation and unemployment.

Fujihara and Mougoué (2007) investigated whether the United States inflation rate was stationary and observed that US inflation had low frequency permanent shocks contrasted with the high frequency permanent shocks.

We analyzed whether the USA inflation is stationary or not. Figure 12.1 shows the consumer price index (CPI) from 1913 to 2011. When we apply the variance ratio test on this data in its entirety we obtain 1.800 indicating that the CPI index is non-stationary in its entirety.

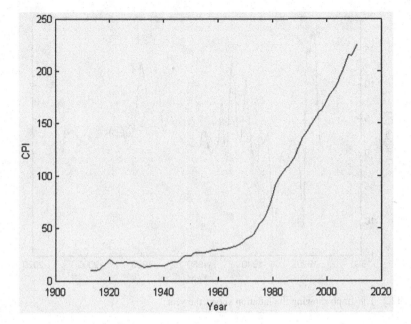

Fig. 12.1 The graph showing the CPI versus the year

The same CPI data can be transformed into inflation by calculating the percentage change of inflation and this data are shown in Fig. 12.2. The variance ratio test results give a variance of 1.047 indicating that inflation is fairly stationary in the period.

The 11 years moving variance ratio is shown in Fig. 12.3. This figure indicates that inflation was stationary at times and non-stationarity at other times.

12.3 Control of Non-stationary Process

In this chapter, we define inflation targeting as a control problem. In the previous section, we concluded that inflation is essentially a non-stationary phenomenon. Given this conclusion, in this section, we will define important aspects that the control of non-stationary phenomenon must have.

Guo and Zhang (2012) applied H_∞ to control vehicle suspension under non-stationary conditions. They observed that the H_∞ control strategy improved results in terms of ride comfort, dynamic suspension deflection, dynamic tire loads, and non-stationary conditions.

Abdoos et al. (2011) applied multi agent Q-learning for traffic light control in non-stationary environments. The results revealed that the technique performed better than the fixed time technique.

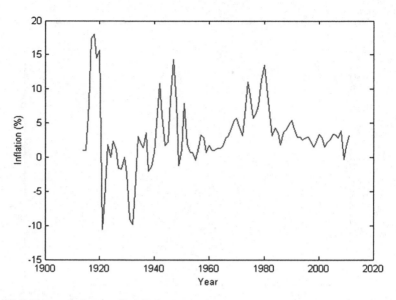

Fig. 12.2 The graph showing the inflation versus the year

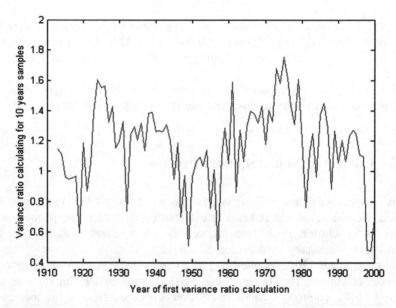

Fig. 12.3 An illustration of the moving variance ratio of inflation calculating over a period of 11 years

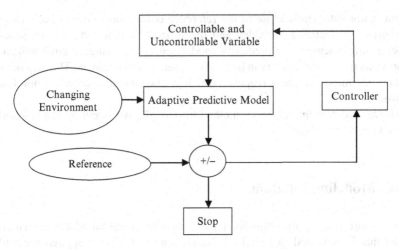

Fig. 12.4 Illustration of the control system for non-stationary environment

Strijbosch et al. (2011) studied the interface between forecasting and stock control for the situation of non-stationary demand. The results obtained contributed to better understanding of operational matters and improved the stock control systems.

Sheng et al. (2011) proposed a system identification and controller design for an active-head slider with non-stationary and non-linear slider dynamics. The proposed system identification technique and controller was observed to give good results.

Köroğlu and Scherer (2011) applied control systems for robust reduction of non-stationary sinusoidal disturbances with measurable frequencies. The proposed method was successfully implemented to control the steering of a ship.

Jasour and Farrokhi (2010) applied neural networks and predictive control and applied these to robotic manipulators in non-stationary environments. The results obtained indicated that the proposed technique was effective.

Qi and Moir (2010) proposed an in-car design to identify the driver's voice and to control in-car infrastructure in a non-stationary noise car environment. They observed that the proposed method was able to identify speech with accuracy of 75 %, as opposed to 10 % benchmark results.

Predictive control systems with non-stationary environment require that the predictive model be dynamic enough to be able to evolve with the predictive model in line with the evolution of non-stationarity. Artificial intelligence has been able to offer models that are sufficiently flexible to be adaptive and evolve. The general control system that is able to deal with non-stationarity is shown in Fig. 12.4. In this figure, the predictive model receives the input data which can be categorized into the controllable and uncontrollable input variables. The predictive model then transforms these input variables into the output. This output is compared to a reference signal, which is the desired output. If the desired predicted output is not sufficiently close to the reference output then a controller identifies appropriate controllable variables and the process is repeated. As this scheme is being implemented, the predictive model senses its changing environment to improve its predictive capacity in line with the sensed environment. The next section explains the multi-layered perceptron, which is adaptable to account for non-linear control.

The next section implements a method for a predictive model, which is a multi-layered neural network.

12.4 Modeling Inflation

In this section we apply multi-layer perceptron to create an adaptive predictive model that is indicated in Fig. 12.4. Nasrzadeh et al. (2011) applied the multi-layer perceptrons and conventional adaptive filters for channel estimation in a code division multiple access (CDMA) system. The results indicated that the multi-layer perceptron gave better results than conventional methods.

Achili et al. (2009) applied adaptive multi-layer perceptron and sliding mode methods for an adaptive parallel robots control. When this technique was implemented on a parallel robot, the technique performed well even in the presence of external disturbances.

Gavrilov et al. (2006) applied multi-layer perceptron and adaptive resonance theory for mobile robots recognition of new objects whereas Shafiq and Moinuddin (2003) applied a multi-layer perceptron for adaptive inverse control and observed that it was able to control a non-minimum phase system and that the inverse multi-layer perceptron was less sensitive to the frequency spectrum of the excitation signal.

Suksmono and Hirose (2003) applied a multi-layer perceptron for adaptive beam forming. The results obtained indicated that the complex-valued back-propagation algorithm performed better than beam forming using a complex-valued least mean square method, the rate of learning convergence, and interferences suppressions.

Langlet et al. (2001) successfully applied a multi-layer perceptron for adaptive pre-distortion for a solid state power amplifier while You and Hong (1996) successfully applied the multi-layer perceptron for blind adaptive equalization, and Riedmiller (1994) successfully applied multi-layer perceptron for some benchmark problem.

The multi-layer perceptron neural network is an input-out model that is able to model data of arbitrary complexity. The network which was discussed in Chap. 3 can be mathematically described as follows (Bishop 1995):

$$y_k = f_{outer}\left(\sum_{j=1}^{M} w_{kj}^{(2)} f_{inner}\left(\sum_{i=1}^{d} w_{ji}^{(1)} x_i + w_{j0}^{(1)}\right) + w_{k0}^{(2)}\right) \qquad (12.4)$$

Here, $w_{ji}^{(1)}$ and $w_{ji}^{(2)}$ are weights in the first and second layer, respectively, from input i to hidden unit j, M is the number of hidden units, d is the number of output units, while $w_{j0}^{(1)}$ and $w_{k0}^{(2)}$ are the weight parameters that represent the biases for the hidden unit j and the output unit k. These weight parameters can be interpreted as instruments that ensure that the model essentially comprehends the data. In this chapter, the parameter $f_{outer}(\bullet)$ is the linear function, while f_{inner} is the hyperbolic tangent function.

The network weights in Eq. 12.4 are identified from the data that contains the input x and input y. There are many techniques that have been applied to identify these network weights, given the observed data. These include the maximum-likelihood as well as Bayesian methods. When the maximum-likelihood framework is applied to identify the network weights, given the observed data, the techniques used include the steepest gradient method, the conjugate method, and the scaled conjugate method (Bishop 1995). The methods that have been applied to identify the network weights, given the observed data, include the Monte Carlo method, the Markov Chain Monte Carlo Method, Simulated Annealing, hybrid Monte Carlo and the shadow hybrid Monte Carlo method (Marwala 2007, 2009, 2010, 2012; Marwala and Lagazio 2011).

The model in Eq. 12.4 can be forced to become adaptive by ensuring that, every time a new observation is sensed from the environment, the network weights are re-estimated by using some optimization procedure. Within the Bayesian framework, this will entail re-estimating the posterior distribution in the light of the observed information.

The model in Eq. 12.4 is applied to model inflation. The input variables are mining output, transport, storage and communication output; financial intermediation, insurance, real estate and business services output; community, social and personal services output; gross value added at basic prices; taxes less subsidies on products; affordability; economic growth; repo rate; GDP growth; household consumption; and investment while the output was the inflation rate of the South African economy. When the multi-layered perceptron was implemented it had 12 inputs, seven hidden nodes, and one output while the activation function in the hidden layer was a hyperbolic tangent, while the activation function in the output layer was linear. To identify the network weights, the maximum-likelihood method was used and the scaled conjugate optimization method was used to find the optimal weights. The results obtained when this procedure was implemented are shown in Fig. 12.5. These results indicate that the multi-layer perceptron is able to model inflation rate.

12.5 Controlling Inflation

Figure 12.4 has a controller which, in this chapter, uses a genetic algorithm. This controller is enabled by minimizing the error between the reference signal and the output signal of the adaptive predictive model in Fig. 12.4. This objective can be

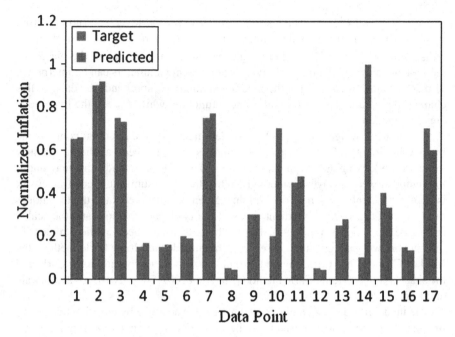

Fig. 12.5 Results obtained when the multi-layer perceptron was used to model the inflation rate

achieved by minimizing the error E and this can be mathematically represented as follows:

$$E = \sum \left(y \left(\left\{ \begin{array}{c} x_u \\ x_c \end{array} \right\} , w \right) - R \right)^2 \tag{12.5}$$

Here, y is the prediction from the adaptive predictive model, w is the weight vector of the trained multi-layer perceptron, x_u is the uncontrolled variables vector, x_c is the controlled variables vector, and R is the reference signal.

Öztürk and Çelik (2012) applied genetic algorithms for speed control of permanent magnet synchronous motors. The results obtained showed that the implemented controller gave better dynamic response than that of the conventional one. Shill et al. (2012) successfully applied quantum genetic algorithms to optimize a fuzzy logic controller, while Takahashi et al. (2012) successfully applied a multi-layer quantum neural network controller designed using a genetic algorithm.

Mahdavian et al. (2012) applied a hybrid genetic algorithm controller for load frequency control of a two-area HVAC/HVDC power system. The results they obtained indicated the effectiveness of the proposed controller on controlling load frequency.

Asan Mohideen et al. (2013) applied a genetic algorithm for system identification and tuning of a modified model reference adaptive controller for a hybrid tank system. The results obtained demonstrated that the genetic algorithm controller gave better transient performance than the PID controller. Furthermore, the proposed method gave good steady-state performance for controlling non-linear processes.

Guo et al. (2012) applied genetic algorithms for designing a steering controller for trajectory tracking of unmanned vehicles. The results obtained demonstrated that this method robustly and accurately track the reference trajectories.

Farouk (2012) applied a genetic algorithm and fuzzy tuning PID controller for a speed control system for marine diesel engines. The results obtained demonstrated that the proposed controller was more effective and gave faster system response than a fuzzy tuned PID controller.

Ghosh and Gude (2012) successfully applied a genetic algorithm tuned optimal controller for glucose regulation in type 1 diabetic subjects. The results obtained showed that the method handled noisy output measurement, modeling error, and delay in sensor measurement well.

Zhang et al. (2012) applied a genetic algorithm fuzzy logic controller for semi-active suspension. The results obtained showed that this method improved the performance of a full car suspension system and was better than passive suspension response.

Valarmathi et al. (2012) applied genetic algorithm controllers and applied this in non-linear liquid tank systems. When a genetic algorithm controller was compared to a Ziegler-Nicholas (ZN) closed loop controller and a ZN open loop technique, it was found to perform better.

Al-Faiz and Sabry (2012) successfully applied a genetic algorithm optimal linear quadratic controller in TCP/AQM router, while Slavov and Roeva (2012) successfully applied a genetic algorithm to tune a PID controller for glucose concentration control.

In this chapter, we apply a genetic algorithm to build a controller shown in Fig. 12.4. This chapter will use game theory to describe genetic algorithms. Game theory is a conceptual technique that is derived on how games are played. It has been applied to economics quite extensively (Wei et al. 2012). Game theory has few key components and these are the players, rules, strategy, and pay-off (Cleveland and Ackleh 2013; Hui and Bao 2013; Baker and Shokrieh 2013; Dev Choudhury and Goswami 2012).

The players in genetic algorithms are the elements of the population. The rules are cross-over, mutation and reproduction. The strategy is the survival of the fittest principle, while the pay-off is the fitness function (Holland 1975; Goldberg 1989, 2002).

A genetic algorithm is randomly initiated by specifying the size of the population of individual solutions. For example, if the population size is 10, then within the context of game theory is a game with ten players. Implementing a genetic algorithm can be viewed as a cooperative game, and this cooperation is enabled through a process called crossover.

The crossover operator is a mechanism which combines genetic information in the population by cutting pairs of chromosomes at random points along their length and exchanging the cut sections over (Banzhaf et al. 1998). This process allows the players in a game to exchange information, thereby, making genetic algorithms a cooperative game.

There are many types of the cross-over methods and here we describe a one crossover point technique. A one point cross-over is conducted by copying a binary string from the beginning of a chromosome to the crossover point from one parent, and the rest is copied from the second parent. For instance, if two chromosomes in binary space a = **1100**1010 and b = **1101**0011 undergo a one-point crossover at the midpoint, then the resulting offspring maybe 11000011 as well as 11011010.

The other operator used in the formulation of a genetic algorithm is called mutation. This is an operator which introduces new information into the player of the game and averts the genetic algorithm simulation from being trapped in a local optimum solution (Goldberg 2002). A simple mutation is applied by randomly choosing a mutation point in the chromosome of each player and randomly inverting it. For instance, if a chromosome in binary space 11001010 is mutated at the second element of the chromosome then the offspring becomes 1**0**001010.

A selection of the proportion of the current population is selected to generate a new population and this is based on the fitness function, which is the pay-off in a genetic algorithm. This selection is attained by applying the fitness-based method, where solutions that are fitter, as measured by Eq. 12.5, have a higher probability of survival. Some selection approaches rank the fitness of each solution and select the best solution, while other techniques rank a randomly selected aspect of the population. There is a significant number of selection processes and, in this chapter we use roulette-wheel selection (Goldberg 2002). Roulette-wheel selection is a genetic operator applied for selecting possible solutions in a genetic algorithm optimization technique. In this technique, each likely process is allotted a fitness function that is applied to map the probability of selection with each individual solution.

This method guarantees that solutions with higher fitness values have higher probabilities of survival than those with a lower fitness value. The advantage of this is that, despite the fact that a solution may have a low fitness value; it may still have some features that are beneficial in the future. The method is recurred until a termination condition has been attained, either for the reason that a selected solution that meets the objective function has been obtained, or for the reason that a stated number of generations have been realized, or the solution has converged or any combination of these.

The neural-network-genetic-algorithm method proposed in this chapter for inflation targeting can be summarized as follows:

- Train a multi-layer perceptron neural network to take variables x and predict the inflation rate y. Here, note that x can be divided into x_c that is controlled and, in this chapter, it is the interest rate and x_u which is not controlled.
- Then create an inflation controller which minimizes, using a genetic algorithm, the distance between the predicted inflation from the neural network and the targeted inflation to identify the required interest rate.

12.6 Experimental Investigation

The inflation rate scheme proposed in this chapter is based on the premise that there is a relationship between the inflation rate and the interest rate. This assumption has been made by several researchers in the past (Kose et al. 2012; Zaheer Butt et al. 2010; Mills 2008; Booth and Ciner 2001; Lanne 2001; Crowder and Hoffman 1996).

This chapter uses Eq. 12.4 to model the relationship between input variables: mining output, transport, storage and communication output; financial intermediation, insurance, real estate and business services output; community, social and personal services output; gross value added at basic prices; taxes less subsidies on products; affordability; economic growth; repo rate; GDP growth; household consumption; as well as investment and output variable inflation rate. These data was obtained from the Reserve Bank of South Africa.

As indicated before, the multi-layered perceptron had 12 inputs, seven hidden nodes and one output, while the activation function in the hidden layer was a hyperbolic tangent and the activation function in the output layer was linear. The network weights were identified using the maximum-likelihood technique and scaled conjugate gradient optimization method.

The data was partitioned into two parts, the 100 examples for creating a multi-layered perceptrons and 15 for inflation targeting. The actual observed inflation rates for these 15 data points were assumed to be the targeted inflation rate. The measure of how well the inflation targeting strategy is how well the targeted inflation rate is to what was actually observed and, at the same time, how close the corresponding interest rate was to the actual one. This is the best we can do this because it is costly to implement this technique in real life because of the associated cost of testing a new method on a real economy.

For the optimization procedure, a genetic algorithm with ten players (population size) was used and simple cross-over, simple mutations, and Roulette wheel reproduction was used. The results obtained when these methods are employees are shown Fig. 12.6.

The results obtained indicate that inflation targeting seems to work. Of course this model can be improved by using more robust predictive model as well as optimization methods.

12.7 Conclusions

In this chapter, an attempt was made to create a system that can be used for inflation targeting. A control system approach was adopted in this regard and this was based on artificial intelligence. The input/output model was constructed using a multi-layered perceptron network and a closed loop control strategy was adopted using a genetic algorithm to control inflation through the manipulation of interest rate. Given the historical inflation rate data, a control scheme was used to determine the

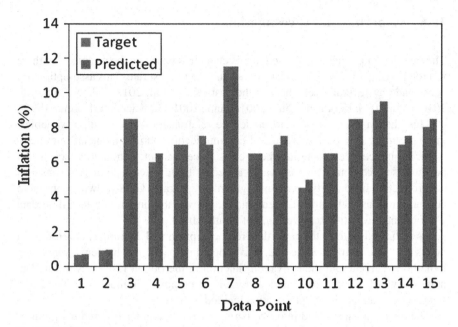

Fig. 12.6 Achieved versus targeted inflation rate

interest rate which was required to attain the given inflation rate. The calculated interest rate was then compared to the historical inflation rate to evaluate the effectiveness of the control strategy and this was found to give good results.

References

Abdoos M, Mozayani N, Bazzan ALC (2011) Traffic light control in non-stationary environments based on multi agent Q-learning. In: Proceedings of the IEEE conference on intelligent transportation systems, Washington, DC, 2011, pp 1580–1585

Abo-Zaid S, Tuzemen D (2012) Inflation targeting: a three-decade perspective. J Policy Model 34:621–645

Achili B, Daachi B, Ali-Cherif A, Amirat Y (2009) Combined multi-layer perceptron neural network and sliding mode technique for parallel robots control: an adaptive approach. In: Proceedings of the international joint conference on neural networks, Atlanta, 2009, pp 28–35

Al-Faiz MZ, Sabry SS (2012) Optimal linear quadratic controller based on genetic algorithm for TCP/AQM router. In: Proceedings of the international conference on future communication networks, Baghdad, 2012, pp 78–83

Amira B, Mouldi D, Feridun M (2013) Growth effects of inflation targeting revisited: empirical evidence from emerging markets. Appl Econ Lett 20:587–591

Asan Mohideen K, Saravanakumar G, Valarmathi K, Devaraj D, Radhakrishnan TK (2013) Real-coded genetic algorithm for system identification and tuning of a modified model reference adaptive controller for a hybrid tank system. Appl Math Model 37(6):3829–3847

Atesoglu HS, Smithin J (2006) Inflation targeting in a simple macroeconomic model. J Post Keynes Econ 28:673–688

Baker M, Shokrieh F (2013) Chip-firing games, potential theory on graphs, and spanning trees. J Comb Theory 120:164–182

Banzhaf W, Nordin P, Keller R, Francone F (1998) Genetic programming – an introduction: on the automatic evolution of computer programs and its applications. Morgan Kaufmann, San Francisco

Bishop CM (1995) Neural networks for pattern recognition. Oxford University Press, Oxford

Bleich D, Fendel R, Rülke J-C (2012) Inflation targeting makes the difference: novel evidence on inflation stabilization. J Int Money Finance 31:1092–1105

Booth GG, Ciner C (2001) The relationship between nominal interest rates and inflation: international evidence. J Multinatl Finance Manage 11:269–280

Broto C (2011) Inflation targeting in Latin America: empirical analysis using GARCH models. Econ Model 28:1424–1434

Carrasco CA, Ferreiro J (2013) Inflation targeting and inflation expectations in Mexico. Appl Econ 45:3295–3304

Cleveland J, Ackleh AS (2013) Evolutionary game theory on measure spaces: well-posedness. Nonlinear Anal Real World Appl 14:785–797

Crowder WJ, Hoffman DL (1996) The long-run relationship between nominal interest rates and inflation: the Fisher equation revisited. J Money Credit Bank 28:102–118

Dev Choudhury NB, Goswami SK (2012) Transmission loss allocation using combined game theory and artificial neural network. Int J Electr Power Energy Syst 43:554–561

Ding H, Kim J (2012) Does inflation targeting matter for PPP? An empirical investigation. Appl Econ Lett 19:1777–1780

Farouk N (2012) Genetic algorithm and fuzzy tuning PID controller applied on speed control system for marine diesel engines. Res J Appl Sci Eng Technol 4:4350–4357

Fujihara RA, Mougoué M (2007) Testing for infrequent permanent shocks: is the US inflation rate stationary? Appl Finance Econ 17:951–960

García-Solanes J, Torrejón-Flores F (2012) Inflation targeting works well in Latin America. Cepal Rev X:37–53

Gavrilov A, Lee Y-K, Lee S (2006) Hybrid neural network model based on multi-layer perceptron and adaptive resonance theory. Lect Notes Comput Sci 3971:707–713

Ghosh S, Gude S (2012) A genetic algorithm tuned optimal controller for glucose regulation in type 1 diabetic subjects. Int J Numer Methods Biomed Eng 28:877–889

Goldberg DE (1989) Genetic algorithms in search, optimization, and machine learning. Addison-Wesley, Reading

Goldberg DE (2002) The design of innovation: lessons from and for competent genetic algorithms. Addison-Wesley, Reading

Granger CWJ, Newbold P (1974) Spurious regressions in econometrics. J Econ 2:111–120

Guo L-X, Zhang L-P (2012) Robust H_∞ control of active vehicle suspension under non-stationary running. J Sound Vib 331:5824–5837

Guo J, Hu P, Li L, Wang R (2012) Design of automatic steering controller for trajectory tracking of unmanned vehicles using genetic algorithms. IEEE Trans Vehicular Technol 61:2913–2924

Harjes T, Ricci LA (2010) A Bayesian-estimated model of inflation targeting in South Africa. IMF Staff Pap 57:407–426

Holland JH (1975) Adaptation in natural and artificial systems. University of Michigan Press, Ann Arbor

Hui ECM, Bao H (2013) The logic behind conflicts in land acquisitions in contemporary China: a framework based upon game theory. Land Use Policy 30:373–380

Jasour AM, Farrokhi M (2010) Control of redundant manipulators in non-stationary environments using neural networks and model predictive control. Proc IFAC 8:8–13

Karahan O, Çolak O (2013) The impact of inflation targeting policy on the inflation uncertainty in Turkey. In: Karasavvoglou A, Polychronidou P (eds) Balkan and Eastern European Countries in the Midst of the Global Economic Crisis, Contrib Econ:49–61, Springer

Kiremire BBE, Marwala T (2008) Non-stationarity detection: a stationarity index approach. In: Proceedings of the IEEE international congress on image and signal process, Hainan, China, pp 373–378

Körolu H, Scherer CW (2011) Scheduled control for robust attenuation of non-stationary sinusoidal disturbances with measurable frequencies. Automatica 47:504–514

Kose N, Emirmahmutoglu F, Aksoy S (2012) The interest rate-inflation relationship under an inflation targeting regime: the case of Turkey. J Asian Econ 23:476–485

Kwiatkowski D, Phillips PCB, Schmidt P, Shin Y (1992) Testing the null hypothesis of stationarity against the alternative of a unit root. J Econom 54:159–178

Langlet F, Abdulkader H, Roviras D, Mallet A, Castanié F (2001) Adaptive predistortion for solid state power amplifier using multi-layer perceptron. In: Proceedings of the IEEE global telecommunications conference, San Antonio, 2001, pp 325–329

Lanne M (2001) Near unit root and the relationship between inflation and interest rates: a reexamination of the Fisher effect. Empir Econ 26:357–366

Levin JH (2004) A model of inflation targeting in an open economy. Int J Financ Econ 9:347–362

Lo AW, MacKinlay AC (1989) The size and power of the variance ratio test. J Econom 40:203–238

Lomax R (2005) Inflation targeting in practice: models, forecasts, and hunches. Atl Econ J 33: 251–265

Lopez C (2009) Euro-zone inflation rates: stationary or regime-wise stationary processes. Econ Bull 29:238–243

Mahdavian M, Wattanapongsakorn N, Azadeh M, Ayati A, Poudeh MB, Jabbari M, Bahadory S (2012) Load frequency control for a two-area HVAC/HVDC power system using hybrid genetic algorithm controller. In: Proceedings of the 9th international conference on electrical engineering/electronics, computer, telecommunications and information technology, Phetchaburi, 2012, art. no. 6254306

Marwala T (2007) Computational intelligence for modelling complex systems. Research India Publications, New Delhi

Marwala T (2009) Computational intelligence for missing data imputation, estimation and management: knowledge optimization techniques. IGI Global Publications, New York

Marwala T (2010) Finite element model updating using computational intelligence techniques. Springer, London

Marwala T (2012) Condition monitoring using computational intelligence methods. Springer, London

Marwala T, Lagazio M (2011) Militarized conflict modeling using computational intelligence techniques. Springer, London

Mills TC (2008) Exploring historical economic relationships: two and a half centuries of British interest rates and inflation. Cliometrica 2:213–228

Nasrzadeh S, Gehasemlou M, Jalali M (2011) Multi-layer perceptrons and conventional adaptive filters for channel estimation in CDMA system. In: Proceedings of the international conference on artificial intelligence, Arizona, USA, pp 458–462

Öztürk N, Çelik E (2012) Speed control of permanent magnet synchronous motors using fuzzy controller based on genetic algorithms. Int J Electr Power Energy Syst 43:889–898

Palma AA, Portugal MS (2009) Empirical analysis of the formation of inflationary expectations in Brazil: an application of artificial neural networks to panel data. Revista de Economia Contemporanea 13:391–438

Perron P (1988) Trends and random walks in macroeconomic time series: further evidence from a new approach. J Econ Dyn Control 12:297–332

Pourroy M (2012) Does exchange rate control improve inflation targeting in emerging economies? Econ Lett 116:448–450

Priestley MB (1988) Non-linear and non-stationary time series analysis. Academic, Waltham

Qi TZ, Moir TJ (2010) Automotive speech control in a non-stationary noisy environment. Int J Comput Appl Technol 39:27–31

Riedmiller M (1994) Advanced supervised learning in multi-layer perceptrons – from backpropagation to adaptive learning algorithms. Comput Stand Interface 16:265–278

Russell B (2011) Non-stationary inflation and panel estimates of United States short and long-run Phillips curves. J Macroecon 33:406–419

Russell B, Banerjee A (2008) The long-run phillips curve and non-stationary inflation. J Macroecon 30:1792–1815

Schwert W (1989) Tests for unit roots: a Monte Carlo investigation. J Bus Econ Stat 7:147–159

Shafiq M, Moinuddin M (2003) Adaptive inverse control using a multi layer perceptron neural network. In: IASTED international conference on modelling identification and control, Innsbruck, 2003, pp 595–599

Sheng G, Huang L, Chang J-Y, He J, Duan S (2011) An approach for non-stationary and nonlinear vibration identification and control of active air-bearing slider system. Microsyst Technol 17:1123–1127

Shill PC, Amin MF, Akhand MAH, Murase K (2012) Optimization of interval type-2 fuzzy logic controller using quantum genetic algorithms. In: IEEE international conference on fuzzy systems, Brisbane, 2012, art. no. 6251207

Slavov T, Roeva O (2012) Application of genetic algorithm to tuning a PID controller for glucose concentration control. WSEAS Trans Syst 11:223–233

Strijbosch LWG, Syntetos AA, Boylan JE, Janssen E (2011) On the interaction between forecasting and stock control: the case of non-stationary demand. Int J Prod Econ 133:470–480

Suksmono AB, Hirose A (2003) Adaptive beamforming by using complex-valued multi layer perceptron. Lect Notes Comput Sci 2714:959–966

Takahashi K, Kurokawa M, Hashimoto M (2012) Remarks on multi-layer quantum neural network controller trained by real-coded genetic algorithm. Lect Notes Comput Sci 7202:50–57

Valarmathi R, Theerthagiri PR, Rakeshkumar S (2012) Design and analysis of genetic algorithm based controllers for nonlinear liquid tank system. In: Proceedings of the IEEE international conference on advances in engineering, science and management, Nagapattinam, 2012, pp 616–620

Wei X, Qu H, Ma E (2012) Decisive mechanism of organizational citizenship behavior in the hotel industry – an application of economic game theory. Int J Hosp Manage 31:1244–1253

Yilmazkuday H (2013) Inflation targeting, flexible exchange rates and inflation convergence. Appl Econ 45:593–603

You C, Hong D (1996) Blind adaptive equalization techniques using the complex multi-layer perceptron. In: Proceedings of the IEEE global telecommunications conference, London, 1996, pp 1340–1344

Zaheer Butt B, Rehman KU, Azeem M (2010) The causal relationship between inflation, interest rate and exchange rate: the case of Pakistan. Transform Bus Econ 9:95–102

Zhang J, Gao R, Zhao Z, Han W (2012) Fuzzy logic controller based genetic algorithm for semi-active suspension. J Sci Ind Res 71:521–527

Chapter 13
Modeling Interstate Conflict: The Role of Economic Interdependency for Maintaining Peace

Abstract This chapter assumes that peace is a necessary condition for healthy economic activities. It explores the role of trade in maintaining peace and, therefore, healthy economic activities. This is done by constructing the relationship between independent variables *Allies*, *Contingency*, *Distance*, *Major Power*, *Capability*, *Democracy*, as well as *Economic Interdependency* and the dependant variable *Interstate Conflict*. The chapter applies artificial intelligence techniques to study the sensitivity of the variable *Economic Interdependency* on driving peace and thus a healthy economic environment.

13.1 Introduction

Any progressive society seeks to build a society that aims to attain the highest form of social, economic, and political advancement of its people. A formula on how to create such a society remains elusive. One important characteristic of a progressive society is that it is a society which is positioned within a state; which is at peace with itself, its neighbors, and the international community. For this reason a progressive society, as a matter of value, should aim for global peace and inspire a culture of the highest form of human development.

Granted that peace is a necessary condition to build a progressive and economically prosperous society, it is consequently important to comprehend the anatomy of interstate conflicts and use this insight to increase the occurrence of peace and economic prosperity. This chapter handles conflict between countries as a scientific phenomenon to be analyzed and comprehended, and then be applied to increase peace. The capacity to scientifically comprehend the causes of militarized interstate conflict, and then to use this knowledge to shape and promote peace in the international context is irrefutably an imperative endeavor for economic prosperity.

In order to comprehend international conflicts, this chapter suggests an artificial intelligence viewpoint to analyze some of the complex behaviors exhibited by interstate conflicts (Lagazio 2001; Marwala and Lagazio 2011a). Artificial intelligence is essentially a method in which computers or machines are capacitated to think like intelligent human beings. Artificial intelligence can be applied to identify conflicts before they occur and, thereby, serve as an early warning system which could be used to prevent destruction of economic and social infrastructure.

As described by Marwala and Lagazio (2011a), *early warning* usually refers to a set of actions whose goal is to gather, integrate, and analyze data with the intention of detecting and identifying the early indicators of an incipient crisis before it bursts out into violence and economic destruction (Alexander 2003). *Conflict prevention*, as an alternative, describes the development and crisis intervention actions intended at reconciling parties with conflicting interests, to prevent the search for different objectives from deteriorating into conflicts of severe intensity (Rupesinghe 1994). The notion of conflict deterrence has also been prolonged to consist of the efforts and management strategies designed and applied to avoid future reversions into violence.

Early warning and conflict prevention are connected and, if they are implemented quickly, can strengthen one another. Early warning is executed for preventive purpose of (Alexander 2003; Marwala and Lagazio 2011a, b):

- Expecting the escalation of violent conflict;
- The development of strategic responses to these crises; and
- To offer choices to decision makers for decision-making and preventive action.

Furthermore, both of them are intended to prevent any type of violent conflict, including war. Early warning and conflict prevention are complex fields with many different approaches, and have a comprehensive number of players involved; including experts, grassroots players, and researchers. Even though a few people will differ with the need for early warning and conflict prevention systems, successful early warning related to conflict prevention has proved to be difficult. There is a requirement to actively engage in crisis prevention, where the first step is the diagnosis and prognosis of *when*, *why*, and *where* conflict will explode in addition to *how* to intervene. This is the same technique as any diagnostic procedure, where the following questions are asked (Marwala and Lagazio 2011a):

(a) What is the issue?
(b) How impending is it?
(c) What are the primary causes?
(d) How do we solve it?
(e) Is the preventive intervention having an effect?

The choices and actions that can be selected and applied are predicated upon a proper understanding of the potential conflict. In essence, early warning becomes one of the conditions for success in conflict prevention.

This chapter emphasizes the relationship between early warning and prevention and how conflict prevention can be improved to spread peace and promote economic activities, and it summarizes the work conducted by Marwala and Lagazio (2011a). The results of all the analyses are integrated into *a controlling model* to offer a workable single solution for increasing peace in the international system of governance. Singular consideration is placed on the three drivers of Kantian peace, democracy, economic interdependence, and international organizations, and how based on the results, the international community should apply these three crucial factors to encourage the diffusion of peace in the international system of governance. It is vital to pay attention to the fact that we are not attempting to analytically deliberate and solve all the conceivable tasks that the international community have difficulty with, when preventing future conflicts.

Hegre et al. (2010) investigated whether trade promoted peace and concluded that it does promote peace, because violence has significant costs. Boehmer (2007) studied the impact of economic crisis, domestic discord, and state efficacy on the decision to initiate interstate conflict, and observed that democracy and economic development offer internal stability and interstate peace. Other researchers who have studied the relationship between trade and peace are Momani (2007), Blanton (2004), and Kanafani (2001).

The objective is to use this improved pockets of understanding that are revealed by artificial intelligence techniques to build practical platforms and solutions for early warning of conflict in order to manage dispute. In this chapter, we apply fuzzy systems, neuro-rough models, optimized rough models, and support vector machines to model interstate conflict. From these models, the role of trade is identified as a critical variable for maintaining peace amongst countries. Furthermore, we apply control systems to control conflict and, again, trade emerges as an important variable.

13.2   The Drivers of Interstate Conflict

In this section we discuss the data and variables that are vital for understanding interstate conflict. The used data set contains a population of politically relevant dyads (pair of countries) from 1885 to 2001, as explained comprehensively by Oneal and Russett (2005), Marwala and Lagazio (2011a, b), Tettey and Marwala (2006a, b), Marwala and Lagazio (2004), Habtemariam et al. (2005), Tettey and Marwala (2007), and Crossingham et al. (2008). These variables are known as dyads because they involve two countries. For instance, the variable distance explains the distance between two countries (Marwala and Lagazio 2004).

The dependent variable of the models consists of a binary variable, which shows the inception of a militarized interstate conflict (Maoz 1999; Marwala and Lagazio 2011b), is called the *peace-conflict-status*. Only dyads with no interstate conflict or with only the initial year of the militarized interstate conflict ranging from any severity to war are included, because our concern, related to early warning, is to estimate the inception of an interstate conflict rather than its perpetuation. For interstate conflict, we apply the conventional definition of conflict, which is a set of interactions between states comprising threats to use military force, demonstration of military force, or definite use of military force (Maoz 1999, 2005; Lagazio and Marwala 2005).

We involve seven dyadic independent variables, and these are *Allies, Contingency, Distance, Major Power, Democracy, Economic Interdependency,* and *Capability. Contingency* is a binary variable, coded 1 if both states share a border and 0 if they do not. *Distance* is a measure of the distance between the two states' capitals. *Major power* is a binary variable coded 1 if either or both states in the dyad are a super power. The variable *Democracy* is measured on a scale where a value of 10 is an extreme democracy and -10 is an extreme autocracy, and we use the value of the less democratic country in the dyad for our analyses. *Economic Interdependency* is measured as the sum of the countries import and export, with its partner divided by the Gross Domestic Product (GDP) of the stronger country, and it measures the level of economic interdependence. *Capability* is the ratio of the total population, plus the number of people in urban areas, plus industrial energy consumption, plus iron and steel production, plus the number of military personnel in active duty, plus military expenditure in the last 5 years, measured from a stronger country to a weaker country. We lag all independent variables by 1 year to make temporally plausible any inference of causation. We then apply artificial intelligence techniques to identify the relationships between the seven independent variables and the interdependent variable *peace-conflict-status*.

13.3   Artificial Intelligence

Artificial Intelligence (AI) is a comparatively recent field which is about constructing intelligent computer systems with the capacity to learn and reason, in the same way that intelligent human beings do (Marwala and Lagazio 2011b). AI has been applied in important disciplines such as making decisions with incomplete information (Marwala 2009), engineering sciences (Marwala 2010), and making robotic machines that take care of the elderly (Kortenkamp et al. 1998).

There are generally two types of AI methods, and these are learning and optimization approaches. Learning techniques imitate the manner in which the human brain functions to construct better computer machines. The second type applies the mechanism of complex social organisms, such as the flock of birds, colony of ants, and schools of fishes to create computers that are capable of adapting without human intervention, by showing dialectic relationships between individual and group intelligence. In this chapter, we apply AI techniques to understand the relationships between dyadic variables and the *peace-conflict-status* between two countries. This is done in order to accomplish the following (Marwala et al. 2009):

- To predict *peace-conflict-status* between two countries given the independent variables *Allies, Contingency, Distance, Major Power, Democracy, Economic Interdependency* and *Capability*;
- To comprehend the factors that drive interstate conflict; and
- To identify strategies and tactics for controlling and managing interstate conflict.

13.3.1 *Support Vector Machines (SVMs) for Classifying Conflicts*

Support vector machines were proposed by Vapnik (1995, 1998) and his co-workers. In this section, we describe support vector machines in the context of conflict classification. Unlike in Chap. 5, where SVMs were applied to perform regression on economic data, in this section we apply SVMs to classify data. The classification problem can be stated as approximating a function $f : R^N \rightarrow \{-1, 1\}$ dependent on input-output training data which are produced from an independently and identically distributed unknown probability distribution $P(\{x\},y)$ in such a way that f is able to classify unseen $(\{x\},y)$ data (Müller et al. 2001; Habtemariam 2006; Marwala and Lagazio 2011a). The function minimizes the expected error and is mathematically expressed as follows (Müller et al. 2001; Habtemariam 2006; Habtemariam et al. 2005; Marwala and Lagazio 2011b):

$$R[f] = \int l\left(f\left(\{x\}\right), y\right) dP\left(\{x\}, y\right) \tag{13.1}$$

where l is a loss function (Müller et al. 2001). For the reason that the fundamental probability distribution P is unknown, Eq. 13.1 is unsolvable directly. To solve this equation, an upper bound for the risk function is identified, and it is given mathematically as follows (Vapnik 1995; Müller et al. 2001; Marwala and Lagazio 2011a):

$$R[f] = R[f]_{emp} + \sqrt{\frac{h\left(\ln\frac{2n}{h} + 1\right) - \ln\left(\frac{\delta}{4}\right)}{n}} \tag{13.2}$$

where $h \in N^+$ is the Vapnik-Chervonenkis (*VC*) dimension of $f \in F$ and $\delta > 0$. The *VC* dimension of a function class F is defined as the biggest number of h coordinates that can be divided in all possible ways by means of functions of that class (Vapnik 1995). The empirical error $R[f]_{emp}$ is a training error given by (Vapnik 1995; Habtemariam 2006; Marwala and Lagazio 2011a):

$$R[f]_{emp} = \frac{1}{n}\sum_{i+1}^{n} l\left(f\left(x_i\right), y_i\right) \tag{13.3}$$

Given that the training sample is linearly separable by a hyper-plane of the form (Vapnik 1998; Habtemariam 2006; Marwala and Lagazio 2011a):

$$f(x) = \langle w, \{x\} \rangle + b \quad with \quad w \in \chi, \quad b \in \Re \tag{13.4}$$

where $\langle ., . \rangle$ is the dot product, $\{w\}$ is an adjustable weight vector and $\{b\}$ is an offset (Müller et al. 2001; Marwala and Lagazio 2011a). The objective of the learning process as pioneered by Vapnik and Lerner (1963) is to identify the hyper-plane with maximum margin of separation from the class of dividing hyper-planes. However, for the reason that practical data normally show complex properties which cannot be divided linearly, more complex classifiers are necessary. To evade the complexity of the non-linear classifiers, the concept of linear classifiers in a feature space can be introduced. SVMs identify linear separating hyper-planes by mapping the input space into a higher dimensional feature space F through substituting x_i by $\phi(x_i)$ to give (Müller et al. 2001; Habtemariam 2006; Marwala and Lagazio 2011a):

$$Y_i\left(\left(\{w\} \cdot \Phi(\{x\}_i)\right) + b\right), i = 1, 2, \ldots, n \tag{13.5}$$

The *VC* dimension h in the feature space F is constrained subject to $h \leq ||W||^2 R^2 + 1$, where R is the radius of the smallest sphere around the training data (Müller et al. 2001). Consequently, minimizing the expected risk can be defined as an optimization problem as follows (Burges 1998; Müller et al. 2001; Schölkopf and Smola 2003; Marwala and Lagazio 2011a):

$$\text{Minimize}\left(\{w\}, b\right) \frac{1}{2}||\{w\}||^2 \tag{13.6}$$

subject to:

$$c_i\left(\{w\}, \{x\}_i - b\right) \geq 1, i = 1, \ldots, n \tag{13.7}$$

Equations 13.6 and 13.7 are the *quadratic programming problem* because they are a problem of optimizing a quadratic function of a number of variables subject to linear constraints (Schölkopf and Smola 2003) and can be expressed as:

$$||\{w\}||^2 = w \cdot w \tag{13.8}$$

$$\{w\} = \sum_{i=0}^{n} \alpha_i c_i \{x\}_i \tag{13.9}$$

It can be shown that the dual of SVM can be expressed in Lagrangian form by maximizing in α_i, (Schölkopf and Smola 2003):

$$L\left(\alpha\right) = \sum_{i=1}^{n} \alpha_i - \frac{1}{2} \sum_{i,j} \alpha_i \alpha_j c_i c_j \{x\}_i^T \{x\}_j$$

$$= \sum_{i=1}^{n} \alpha_i - \frac{1}{2} \sum_{i,j} \alpha_i \alpha_j c_i c_j k\left(\{x\}_i, \{x\}_j\right), i = 1, \ldots, n \tag{13.10}$$

subject to:

$$\alpha_i \geq 0, i = 1, \ldots, n \qquad (13.11)$$

and subject to the following constraints:

$$\sum_{i=1}^{n} \alpha_i c_i = 0 \qquad (13.12)$$

Here, the kernel is (Müller et al. 2001):

$$k\left(\{x\}_i, \{x\}_j\right) = \{x\}_i \cdot \{x\}_j \qquad (13.13)$$

13.3.1.1 Soft Margin

An improved maximum margin idea that enables misclassified data points was proposed by Cortes and Vapnik (1995). In the absence of a hyper-plane that can distinguish between a "yes" or "no" data points, the *Soft Margin* method chooses a hyper-plane that separates data points effectively while maximizing the distance to the nearest data points. The method introduces slack variables, γ_i which quantify the extent of misclassification of data points, and can be expressed as follows (Cortes and Vapnik 1995):

$$c_i\left(\{w\} \cdot \{x\}_i - b\right) \geq 1 - \gamma_i, 1 \leq i \leq n \qquad (13.14)$$

In order to penalize non-zero γ_i augments, the objective to ensure a trade-off between margin and error penalty, a function needs to be identified. By assuming a linear penalty function, the optimization problem can be expressed by minimizing $\{w\}$ and γ_i of the objective function (Cortes and Vapnik 1995):

$$\frac{1}{2}\|\{w\}\|^2 + C\sum_{i=1}^{n} \gamma_i \qquad (13.15)$$

subject to:

$$c_i\left(\{w\} \cdot \{x\}_i - b\right) \geq 1 - \gamma_i, \gamma_i \geq 0, i = 1, \ldots, n \qquad (13.16)$$

where C is the capacity. These Eqs. 13.15 and 13.16, can be expressed in Lagrangian form by optimizing in terms of $\{w\}$, γ, b, α and β of the following equation (Cortes and Vapnik 1995):

$$\min_{\{w\},\gamma,b} \max_{\alpha,\beta}$$

$$\left\{ \frac{1}{2}\|\{w\}\|^2 + C \sum_{i=1}^{n}\gamma_i - \sum_{i=1}^{n}\alpha_i \left[c_i \left(\{w\}\cdot\{x\}_i - b\right) - 1 + \gamma_i\right] - \sum_{i=1}^{n}\beta_i\gamma_i \right\} \tag{13.17}$$

here $\alpha_i, \beta_i \geq 0$. The advantage of a linear penalty function is that the slack variables are eliminated from the dual problem and, as a result, C is merely a redundant constraint on the Lagrange multipliers. The application of non-linear penalty functions to decrease the effect of outliers causes the optimization problem non-convex and difficult solve.

13.3.1.2 Non-linear Classification

The kernel function was applied to transform the linear SVM procedure into non-linear classifiers (Aizerman et al. 1964; Boser et al. 1992) by substituting the dot product by a non-linear kernel function to fit the maximum-margin hyper-plane. The dot product transformation is non-linear and the transformed space is in high dimensions and, as an example, a Gaussian radial basis function kernel transforms the feature space into a Hilbert space of infinite dimension. In this section, we apply the radial basis function kernel which can be written as follows (Vapnik 1995; Müller et al. 2001):

$$k\left(\{x\}_i, \{x\}_j\right) = \exp\left(-\gamma\|\{x\}_i - \{x\}_j\|^2\right), \gamma > 0 \tag{13.18}$$

To identify the variables of the maximum-margin hyper-plane, the optimization method can be used to solve the objective function using an interior point technique based on the Karush-Kuhn-Tucker (KKT) conditions (Kuhn and Tucker 1951; Karush 1939). The KKT method is a generalized type of the Lagrangian method.

13.3.2 Fuzzy Sets for Classifying Conflicts

Neuro-fuzzy method combines neural networks and fuzzy logic (Jang 1993; Jang et al. 1997). It is a system that combines the human-like reasoning characteristics of fuzzy systems with the learning characteristics of neural systems resulting in a universal approximator with comprehendible IF-THEN rules. As explained by Marwala and Lagazio (2011a), neuro-fuzzy systems implicate two conflicting characteristics: interpretability versus accuracy. The neuro-fuzzy in fuzzy modeling research consists of linguistic fuzzy modeling that is more interpretable (Mamdani 1974) and accurate fuzzy modeling focused (Sugeno and Kang 1988; Sugeno 1985; Takagi and Sugeno 1985).

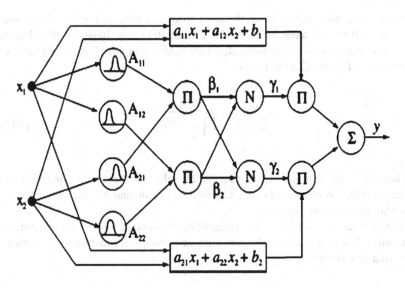

Fig. 13.1 An example of a two-input first order Takagi-Sugeno fuzzy model

Fuzzy logic notions offer a technique of expressing vague models of reasoning. It bears a similarity to human reasoning in its application of estimated information and uncertainty to produce decisions. The estimation of the output is accomplished by a calculating structure known as the fuzzy inference system. The fuzzy inference system transforms fuzzy inputs to the output. This fuzzy inference system implements the inputs using fuzzy set theory, fuzzy *if-then* rules, and fuzzy reasoning to attain the output. It includes the fuzzification of the input variables, assessment of rules, aggregation of the rule outputs, and then the de-fuzzification (*i.e.* extraction of a crisp value which best represents a fuzzy set) of the result.

In this chapter, we apply the Takagi-Sugeno neuro-fuzzy model and this is shown in Fig. 13.1 (Babuska and Verbruggen 2003). In this model, the antecedent part of the rule is a fuzzy proposition and the consequent function is an affine linear function of the input variables and is mathematically defined as follows (Takagi and Sugeno 1985):

$$R_i : \text{If } x \text{ is } A_i \text{ then } y_i = a_i^T x + b_i \qquad (13.19)$$

where R_i is the ith fuzzy rule, x is the input vector, A_i is a fuzzy set, a_i is the consequence parameter vector, b_i is a scalar offset and $i = 1, 2, \ldots, K$. The parameter K is the number of rules in the fuzzy model. Too few rules in the fuzzy model comes at a cost of accuracy, while too many rules leads to a complex model with redundant fuzzy rules compromising the integrity of the model (Sentes et al. 1998).

On implementing this method, the input space is partitioned into the input space in order to form the antecedents of the fuzzy rules. The shapes of the Gaussian membership functions of the antecedents are selected in this chapter and these can be written as follows (Zadeh 1965).

$$\mu^i(x) = \prod_{j=1}^{n} e^{-\frac{\left(x_j - c^i_j\right)^2}{\left(b^i_j\right)^2}} \qquad (13.20)$$

Here, μ^i is the antecedent value for the ith rule, n is the number of antecedents of the ith rule, c is the center of the Gaussian function, and b is the variance of the Gaussian membership function.

The consequent function in the Takagi-Sugeno neuro-fuzzy model can either be constant or linear and, in this chapter, a linear consequent function is implemented (Babuska and Verbruggen 2003):

$$y_i = \sum_{j=1}^{n} p_{ij} x_j + p_{i0} \qquad (13.21)$$

where p_{ij} is the jth parameter of the ith fuzzy rule. When $y_i = p_i$, the Takagi-Sugeno neuro-fuzzy model is then a Mamdani inference system (Mamdani 1974). The output y of the inference system is calculated by a weighted average of the individual rules' contributions as (Babuska and Verbruggen 2003):

$$y = \frac{\sum_{i=1}^{K} \beta_i(x) y_i}{\sum_{i=1}^{K} \beta_i(x)} = \frac{\sum_{i=1}^{K} \beta_i(x) \left(a_i^T x + b_i\right)}{\sum_{i=1}^{K} \beta_i(x)} \qquad (13.22)$$

where $\beta_i(x)$ is the activation of the ith rule and parameters a_i then approximate models of the non-linear system under consideration (Babuska and Verbruggen 2003). There are two techniques to train the neuro-fuzzy method (Babuska and Verbruggen 2003):

1. Fuzzy rules can be extracted from expert knowledge and used to create an initial model.
2. The number of rules can be identified from gathered data using a model selection procedure.

13.3.3 Neuro-Rough Sets for Classifying Conflicts

Bayesian rough sets have been studied extensively recently and Li et al. (2010) applied a probabilistic rough set model with variable precision on Bayesian decisions and observed a reduction of the error risk. The Bayesian framework can be expressed as (Marwala 2007a, b; Marwala and Crossingham 2008; Marwala and Lagazio 2011a):

$$P(M|D) = \frac{P(D|M)p(M)}{p(D)} \tag{13.23}$$

where $M = \begin{Bmatrix} w \\ G_x \\ N_r \\ R \end{Bmatrix}$

$P(M|D)$ is the probability of the rough set model given the observed data, $P(D|M)$ is the probability of the data given the assumed rough set model, $P(M)$ is the prior probability of the rough set model, and $P(D)$ is the probability of the data. The probability of the data given the assumed rough set model and the resulting error may be approximated as follows (Marwala and Crossingham 2008; Marwala and Lagazio 2011b):

$$P(D|M) = \frac{1}{z_1} \exp(-error) = \frac{1}{z_1} \exp\{A(w, N_r, R, G_x) - 1\} \tag{13.24}$$

$$error = \sum_l^L \sum_k^K \left(t_{lk} - \left(f_{outer} \left(\sum_{j=0}^M \gamma_{kj}(G_x, R)_{kj} f_{inner} \left(\sum_{i=0}^d w_{ji}^{(1)} x_i \right) \right) \right)_{lk} \right)^2 \tag{13.25}$$

where z_1 is the normalization constant, L is the number of outputs, and K is the number of training examples. In this problem, the prior probability is linked to the concept of reduct. It is the prior knowledge that the best rough set model is the one with the minimum number of rules (N_r), and that the best network is the one whose weights are of the same order of magnitude. Therefore, the prior probability may be written as follows (Marwala 2007b; Marwala and Crossingham 2008; Marwala and Lagazio 2011b):

$$P(M) = \frac{1}{z_2} \exp\left\{-\alpha N_r - \beta \sum w^2\right\} \tag{13.26}$$

where z_2 is the normalization constant and β is the hyper-parameter of the network weights. Given the observed data, the posterior probability of the model is thus (Marwala and Crossingham 2008; Marwala and Lagazio 2011b):

$$P(M|D) = \frac{1}{z} \exp \left\{ A(N_r, R, G_x) - 1 - \alpha N_r - \beta \sum w^2 \right\} \qquad (13.27)$$

where z is the normalization constant and α is the hyper-parameter of the number of rules. Since the number of rules and the rules generated, given the data, depends on the nature of granulization, we shall sample in the granule space as well as the network weights using a procedure called Markov Chain Monte Carlo (MCMC) simulation (Bishop 1995).

The way in which the probability distribution in Eq. 13.28 is sampled, is to randomly produce a succession of granule-weight vectors and accepting or rejecting them based on how probable they are, using the Metropolis algorithm (Metropolis et al. 1953). The MCMC produces a chain of granules and network weights and accepts or rejects them using the Metropolis algorithm. The application of the Bayesian approach and MCMC neuro-rough sets give a probability distribution function of the granules and network weights and thus the distribution of the neuro-rough model outputs. From these distribution functions, the average approximation of the neuro-rough set model and the variance of that approximation can be computed. Applying the laws of probability theory gives the following distribution of the output parameter y (Marwala 2007a, b; Marwala and Crossingham 2008; Marwala and Lagazio 2011a):

$$p(y|x, D) = \int p(y|x, M)p(M|D)dM \qquad (13.28)$$

Equation 13.28 cannot be solved analytically because of the high dimension of the granule and weight space. For this reason, the solution of this equation may be estimated as (Marwala 2007a, b; Marwala and Crossingham 2008; Marwala and Lagazio 2011a):

$$\tilde{y} \cong \frac{1}{L} \sum_{i=I}^{Z+L-1} F(M_i) \qquad (13.29)$$

F is a model that gives the output given the input, is the average prediction of the Bayesian neuro-rough set model (M_i), Z is the number of initial states that are discarded in the hope of reaching a stationary posterior distribution function disregarded, and L is the number of retained samples. The MCMC technique was applied by sampling a stochastic process consisting of random variables $\{gw_1, gw_2, \ldots, gw_n\}$ by introducing random changes to granule-weight vector $\{gw\}$ and either accepting or rejecting the state using the Metropolis algorithm. The Metropolis algorithm uses the differences of posterior probabilities between two states as follows (Metropolis et al. 1953; Marwala and Crossingham 2008).

$$\text{If } P\left(M_{n+1}|D\right) > P\left(M_n|D\right) \text{ then accept } M_{n+1}, \qquad (13.30)$$

$$\text{Else accept if } \frac{P\left(M_{n+1}|D\right)}{P\left(M_n|D\right)} > \xi \text{ where } \xi \in [0,1] \qquad (13.31)$$

else reject and randomly generate another model M_{n+1}.

13.3.4   Automatic Relevance Determination (ARD)

It is useful for the relationship that exists between variables to be identified. In this chapter, these are the seven variables and the conflict status. In order to achieve this task, the automatic relevance determination (ARD), which is based on the multi-layered perceptron method, is used (MacKay 1991, 1992). The ARD framework is implemented by introducing a hyper-parameter that is associated with each variable in the prior distribution of the Bayesian formulation of the training of the multi-layer perceptron network (Bishop 1995).

The ARD method that is based on the multi-layer perceptron is implemented estimating the α_k^{MP} (hyper-parameter associated with the kth input variable), β^{MP} (hyper-parameter associated with the error between the model prediction and the target data) and the most Probable weight, $\{w\}^{MP}$ as follows (MacKay 1991; Bishop 1995):

1. Randomly select the initial values for the hyper-parameters.
2. Estimate $\{w\}^{MP}$ by training the multi-layer perceptron neural network using the scaled conjugate gradient algorithm to minimize the following cost function:

$$E = \beta E_D + \alpha E_W$$

$$= -\beta \sum_{n=1}^{N} \sum_{k=1}^{K} \{t_{nk} \ln(y_{nk}) + (1 - t_{nk}) \ln(1 - y_{nk})\} - \frac{\alpha}{2} \sum_{j=1}^{W} w_j^2$$

where N and K are the number of outputs and training data, respectively, and y and t are the predicted and target output, respectively.
3. Apply the evidence framework to estimate the hyper-parameters using:

$$\beta^{MP} = \frac{N - \gamma}{2E_D\left(\{w\}^{MP}\right)} \text{ and } \alpha_k^{MP} = \frac{\gamma_k}{2E_{W_k}\left(\{w\}^{MP}\right)}$$

where $\gamma = \sum_k \gamma_k, 2E_{W_k} = \{w\}^T [I_k]\{w\}$ and $\gamma_k = \sum_j \left(\frac{\pi_j - \alpha_k}{\eta_j} \left([V]^T [I_k][V]\right)_{jj} \right)$
whereas η_j and $[V]$ are the eigenvalues and eigenvectors of $[A]$.
4. Repeat Steps 2 and 3 until convergence.

13.4   Controlling Interstate Conflict

There is a revolutionary statement that says: *"The aim of a revolutionary is not merely to understand the world but to actually change it"*. Contextualizing this famous statement, the aim of using AI models in studying interstate conflict is not merely to predict the onset of conflict but actually to use the prediction to prevent conflict and thus promote economic activities. In this chapter, AI is used to relate the prediction of conflict and identification of causes of conflict with the choice of the correct preventive action. The AI model is applied to control conflicts and develop a proper scientific understanding on how to select the right preventive action.

A control system is essentially a procedure where the input of a system is changed to attain a desired outcome. To attain this, a model that characterizes the relationship between the input and the outcome are to be obtained. In this chapter, this model involves characterizing the relationship between the militarized interstate dispute dyadic variables and the *peace-conflict* status. A number of methods can be used to identify such relationships. In this chapter, a Bayesian neural network trained using the hybrid Monte Carlo method is used to identify such a relationship and more details on this method can be found in Bishop (1995), as well as Marwala and Lagazio (2011a). The model that predicts the *peace-conflict* status given the militarized interstate dispute dyadic variables, the next task is to apply this method to identify the set of variables that ensure that conflict can be controlled by reducing the incidence of conflict. The justification for the development of the interstate dispute prediction infrastructure is to maximize the incidence of peace and minimize the incidence of conflict. This is attained by control theory to conflict resolution.

The control system used in this chapter is shown in Fig. 13.2 and this confirms that it has two modules (Marwala and Lagazio 2011a):

- The Bayesian neural network that was applied to identify the relationship between the militarized interstate dispute dyadic variables and the *peace-conflict* status; and
- The optimization element that evaluates the difference between the *peace-conflict* and the preferred output, which is peace, and categorizes the set of inputs that minimize the distance between the predicted output (from the Bayesian neural network) and the desired output (*peace*).

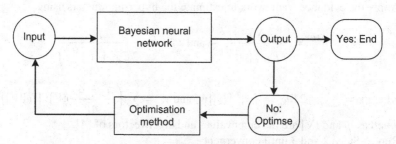

Fig. 13.2 Feedback control loop that uses Bayesian neural networks and an optimization method

The Bayesian network component which takes the militarized interstate dispute dyadic variables as input vector $\{x\}$ given the network weight vector $\{w\}$ and predicts the *peace-conflict* status as output scalar y can be mathematically written as follows (Marwala and Lagazio 2011a):

$$y = f(\{x\}, \{w\}) \tag{13.32}$$

The network weights are identified using the learning process, which is through the Bayesian framework and the details are in Bishop (1995) and Marwala and Lagazio (2011a).

The second element of the control loop is to identify the input given the desired output and an optimization method is used. The objective function of the optimization problem is (Marwala and Lagazio 2011a):

$$error = \sum (y - t_d)^2 \tag{13.33}$$

Here, y is the Bayesian neural network output and t_d is the desired target output. Equation 13.33 is solved using the golden section search technique and simulated annealing (Marwala and Lagazio 2011a).

13.5 Investigation and Results

The correlates of war (COW) data are used to generate training and testing sets (Anonymous 2012). More information on this data set can be found in (Marwala and Lagazio 2011a). The training data set consists of 500 conflict- and 500 non-conflict cases, and the test data consists of 392 conflict data and 392 peace data. A balanced training set, with a randomly selected equal number of conflict- and non-conflict cases was chosen to yield robust classification and stronger comprehensions on the explanation of conflicts. The data were normalized to fall between 0 and 1.

Support vector machines, Takagi-Sugeno neuro-fuzzy systems, and the Bayesian neuro-rough sets were implemented to model militarized interstate dispute data and the results obtained are shown in Table 13.1.

The results obtained demonstrate that the three methods gave similar results on predicting the *peace-conflict* status. The ARD framework was then used to rank variables with regard to their influence on the militarized interstate dispute.

Table 13.1 Classification results

Method	True conflicts TC (%)	True peace TP (%)
Takagi-Sugeno neuro-fuzzy	77	73
Support vector machine	76	74
Bayesian rough set model	76	75

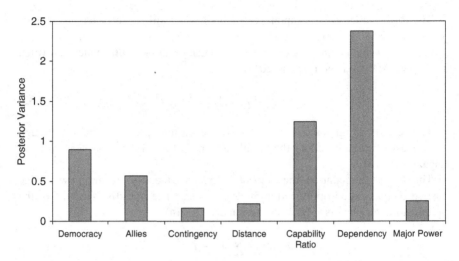

Fig. 13.3 Relevance of each variable with regards to the classification of MIDs (Marwala and Lagazio 2011a)

The ARD was implemented, the hyper-parameters computed, and then the inverse of the hyper-parameters was computed, and the results are demonstrated in Fig. 13.3. Figure 13.3 shows that *Economic Interdependency* has the highest influence on *peace-conflict* status, followed by *Capability*, followed by *Democracy*, and then *Allies*. The remaining three variables, that is, *Contingency*, *Distance*, and *Major Power*, have similar impact although it is much smaller in comparison with the other two variables, democracy and economic interdependence, and the two realist variables, allies and difference in capabilities.

The results in Fig. 13.3 indicate that all the variables used in this book influence the conflict and peace outcome. However, alliance and power ratio play a part in providing opportunities and incentives for interstate action and, therefore, they also have an important effect on promoting peace or conflict between states. Overall, the results first support the theory of democratic peace, which claims that democracies never go to war. Secondly, the results indicate the importance of economic interdependence and economic ties for promoting peace. In addition, this result underlines that the relationship between peace and the Kantian factors is not bi-directional. Economic interdependence interacts with democracy to enhance its own influence, as well as democracy's influence on peace.

However, the three remaining realist variables, *Distance*, *Contingency*, and *Major Power*, cannot be completely ignored. They still provide the ex-ante conditions for war to happen. For example, Swaziland and Bahamas have a lower probability of going to war, primarily because they are so far apart and, as a result, they have no incentives to go to war. The same cannot be said in relation to major powers. Great powers have the capacity to engage in distant conflicts as well as the incentives to do so. In summary, the constrains that the high level of economic interdependence, democracy, allies, as well as difference in capability ratio exert on interstate behavior, are activated when the opportunities for conflicts, provided by geographical proximity and/or great power interests, are in place.

When control strategies are applied to identify the vital independent variables for maximizing peace, it is observed that all four controllable dyadic variables i.e. *Allies, Democracy, Economic Interdependency*, and *Capability* could be used concurrently to avoid all the correctly predicted interstate conflicts. In addition, it was observed that, either *Economic Interdependency* or *Capability* could also be used to avoid all the correctly predicted conflicts, followed by controlling only *Democracy* and then controlling *Allies*. In conclusion, by comparing outcomes from a single and multiple approaches, it is observed that *Economic Interdependency* and *Democracy* are crucial variables to realize peace because, even in a multiple approach, they necessitate substantial changes, which are close to their single level requirement in comparison to the other dyadic variables. This implies that, substantial changes in *Economic Interdependency* and *Democracy* are essential, even if the other dyadic variables have been positively influenced to attain peace. It is worth noting that the techniques applied in this chapter can guide prevention policy or strategies, but should not be used in isolation. A case-by-case method requires to be combined with the results of the controlling model. It would not be worthwhile to apply this AI model thoughtlessly without supporting the result with contextual information.

13.6 Conclusions

Contemporary advances in the interstate conflict literature have underscored the significance of treating international conflicts as complex phenomena, showing complex interactions amongst the relevant militarized interstate conflict variables. Persistently, the relationships between the characteristics of a pair of states and the probability of militarized interstate conflicts have been demonstrated to be reliable in both time and space. The interstate characteristics that have been observed to affect the onset of militarized interstate conflicts are *Economic Interdependence, Democracy, Distance, Relative Power* and *Alliances*. If two states are both reputable democracies, the probability of them engaging in war is almost non-existent, and this is what political scientists call the *Theory of Democratic Peace*. For this reason, the likelihood of, for example, the United States and South Africa ever going to war is very small. Likewise, if the two states are economically interdependent, the probability of them engaging in war is low. For example, Canada and the United States are highly unlikely ever going to war. Therefore, the variables *Democracy* and *Economic Interdependence* offer essential constraints on a state's behaviour and intention to wage a war. Additionally, if two states' capitals are positioned *close* together, the probability of them engaging in war if there is a dispute is high. For this reason, countries like Swaziland and Bahamas are highly unlikely to go to war. In addition, if one of the states is a *superpower*, the distance between the two states' capitals becomes immaterial, because the capability of a superpower to fight a distant war is high. If the difference in power between two states is low, their willingness to use force will also be low, since equal power works as a deterrent. Lastly, the number of *standing alliances* also affects the probability of militarized interstate conflicts, with more alliances increasing the probability of peace. Alliances are constraining the probability of war with non-allies, providing a deterrence mechanism, similar to relative power, while they also reduce the probability of war among their members. *Distance, Relative Power* and *Alliances* provide the state with the opportunities to wage a war. The constraints which are enforced by *Democracy* and *Economic Interdependence*, and the opportunities offered by *Distance, Relative Power* and *Alliances* interact among each other and generate different routes to war and peace. This analysis demonstrates that there is a major effect between *Economic Interdependence, Democracy* and *peace-conflict-status*. Nevertheless, as an alternative to exerting a constant effect, *Economic Interdependence* and *Democracy* vary their effect as they are either facilitated or not facilitated by interaction between themselves and other variables.

References

Aizerman M, Braverman E, Rozonoer L (1964) Theoretical foundations of the potential function method in pattern recognition learning. Autom Remote Control 25:821–837

Alexander A (2003) Early warning and the field: a cargo cult science? Berghof Research Center for Constructive Conflict Management, Berlin

Anonymous (2012) Correlates of war project. http://www.correlatesofwar.org/. Last accessed 20 Sept 2012

Babuska R, Verbruggen H (2003) Neuro-fuzzy methods for nonlinear system identification. Annu Rev Control 27:73–85

Bishop CM (1995) Neural networks for pattern recognition. Oxford University Press, Oxford

Blanton RG (2004) The liberal illusion: does trade promote peace? J Polit 66:660–661

Boehmer C (2007) The effects of economic crisis, domestic discord, and state efficacy on the decision to initiate interstate conflict. Polit Policy 35:774–809

Boser BE, Guyon IM, Vapnik VN (1992) A training algorithm for optimal margin classifiers. In: Haussler D (ed) 5th annual ACM workshop on COLT. ACM Press, Pittsburgh

Burges C (1998) A tutorial on support vector machines for pattern recognition. Data Mining Knowl Discov 2:121–167

Cortes C, Vapnik V (1995) Support-vector networks. Mach Learn 20:273–297

Crossingham B, Marwala T, Lagazio M (2008) Optimized rough sets for modelling interstate conflict. In: Proceedings of the IEEE international conference on man, systems and cybernetics, Singapore, 2008, pp 1198–1204

Habtemariam E (2006) Artificial intelligence for conflict management. Master thesis, University of the Witwatersrand, Johannesburg

Habtemariam E, Marwala T, Lagazio M (2005) Artificial intelligence for conflict management. In: Proceedings of the IEEE international joint conference on neural networks, Montreal, 2005, pp 2583–2588

Hegre H, Oneal JR, Russett B (2010) Trade does promote peace: new simultaneous estimates of the reciprocal effects of trade and conflict. J Peace Res 47:763–774

Jang JSR (1993) ANFIS: adaptive-network-based fuzzy inference system. IEEE Trans Syst Man Cybern 23:665–685

Jang JSR, Sun CT, Mizutani E (1997) Neuro-fuzzy and soft computing: a computational approach to learning and machine intelligence. Prentice Hall, Toronto

Kanafani N (2001) Trade – a catalyst for peace? Econ J 111:F276–F290

Karush W (1939) Minima of functions of several variables with inequalities as side constraints. M.Sc. thesis, University of Chicago, Chicago

Kortenkamp D, Bonasso R, Murphy R (1998) Artificial intelligence and mobile robots. MIT Press, Cambridge, MA

Kuhn HW, Tucker AW (1951) Nonlinear programming. In: Proceedings of 2nd Berkeley symposium, University of California Press, Berkeley, pp 481–492

Lagazio M (2001) An early warning information system for militarised interstate conflicts: combining the interactive liberal peace proposition with neural network modeling. Ph.D. thesis, University of Nottingham, Nottingham

Lagazio M, Marwala T (2005) Assessing different bayesian neural network models for militarized interstate dispute. Soc Sci Comput Rev 24:1–12

Li L, Wang J, Jiang J (2010) Bayesian decision model based on probabilistic rough set with variable precision. Commun Comput Inf Sci 105:32–39

MacKay DJC (1991) Bayesian methods for adaptive models. Ph.D. thesis, California Institute of Technology, Pasadena

MacKay DJC (1992) A practical bayesian framework for back propagation networks. Neural Comput 4:448–472

Mamdani EH (1974) Application of fuzzy algorithms for the control of a dynamic plant. Proc IEEE 121:1585–1588

Maoz Z (1999) Dyadic militarized interstate disputes (DYMID1.1) dataset-version 1.1. ftp://spirit. tau.ac.il./zeevmaoz/dyadmid60.xls. Password protected. Last accessed Aug 2000

Maoz Z (2005) Dyadic MID Dataset (version 2.0): http://psfaculty.ucdavis.edu/zmaoz/dyadmid. html. Last Accessed 7 Mar 2013

Marwala T (2007a) Bayesian training of neural network using genetic programming. Pattern Recognit Lett 28:1452–1458

Marwala T (2007b) Computational intelligence for modelling complex systems. Research India Publications, Delhi

Marwala T (2009) Computational intelligence for missing data imputation, estimation and management: knowledge optimization techniques. IGI Global Publications, New York

Marwala T (2010) Finite element model updating using computational intelligence techniques. Springer, London

Marwala T, Crossingham B (2008) Neuro-rough models for modelling HIV. In: Proceedings of the IEEE international conference on systems, man, and cybernetics, Singapore, 2008, pp 3089–3095

Marwala T, Lagazio M (2004) Modelling and controlling interstate conflict. In: Proceedings of the IEEE international joint conference on neural networks, Budapest, 2004, pp 1233–1238

Marwala T, Lagazio M (2011a) Militarized conflict modeling using computational intelligence. Springer, London

Marwala T, Lagazio M (2011b) The anatomy of interstate conflicts: an artificial intelligence perspective. Thinker 30:40–42

Marwala T, Lagazio M, Tettey T (2009) An integrated human-computer system for controlling interstate disputes. Int J Comput Appl 31(4):239–246. doi:10.2316/Journal.202.2009.4.202-2410

Metropolis N, Rosenbluth AW, Rosenbluth MN, Teller AH, Teller E (1953) Equations of state calculations by fast computing machines. J Chem Phys 21:1087–1092

Momani B (2007) A middle east free trade area: economic interdependence and peace considered. World Econ 30:1682–1700

Müller KR, Mika S, Ratsch G, Tsuda K, Scholkopf B (2001) An introduction to kernel-based learning algorithms. IEEE Trans Neural Netw 12:181–201

Oneal J, Russett B (2005) Rule of three, let it be? When more really is better. Confl Manag Peace Sci 22:293–310

Rupesinghe K (1994) Early warning and preventive diplomacy. J Ethno-Dev 4:88–97

Schölkopf B, Smola AJ (2003) A short introduction to learning with kernels. In: Mendelson S, Smola AJ (eds) Proceedings of the machine learning summer school. Springer, Berlin

Sentes M, Babuska R, Kaymak U, van Nauta LH (1998) Similarity measures in fuzzy rule base simplification. IEEE Trans Syst Man Cybern B Cybern 28:376–386

Sugeno M (1985) Industrial applications of fuzzy control. Elsevier Science Publication Company, Amsterdam

Sugeno M, Kang G (1988) Structure identification of fuzzy model. Fuzzy Sets Syst 28:15–33

Takagi T, Sugeno M (1985) Fuzzy identification of systems and its applications to modeling and control. IEEE Trans Syst Man Cybern 15:116–132

Tettey T, Marwala T (2006a) Neuro-fuzzy modeling and fuzzy rule extraction applied to conflict management. Lect Notes Comput Sci 4234:1087–1094

Tettey T, Marwala T (2006b) Controlling interstate conflict using Neuro-fuzzy modeling and genetic algorithms. In: Proceedings of the 10th IEEE international conference on intelligent engineering systems, London, 2006, pp 30–44

Tettey T, Marwala T (2007) Conflict modelling and knowledge extraction using computational intelligence methods. In: Proceedings of the 11th IEEE international conference on intelligent engineering systems, Budapest, 2007, pp 161–166

Vapnik V (1995) The nature of statistical learning theory. Springer, New York

Vapnik V (1998) Statistical learning theory. Wiley-Interscience, New York

Vapnik V, Lerner A (1963) Pattern recognition using generalized portrait method. Autom Remote Control 24:774–780

Zadeh LA (1965) Fuzzy sets. Inf Control 8:338–353

Chapter 14
Conclusions and Further Work

Abstract This chapter summarises this book and makes recommendations for further studies. It demonstrates that indeed artificial intelligence is a viable tool for analyzing economic data. It also demonstrates that the accuracy of the artificial intelligence method depends on the problem at hand and that there is a wide scope of applying other emerging artificial intelligence techniques to model economic data.

14.1 Conclusions

This book introduced economic modeling based on artificial intelligence techniques. The artificial intelligence methods used included multi-layer perceptrons, radial basis functions, support vector machines, rough sets techniques, automatic relevance determination, autoassociative network, particle swarm optimization, genetic algorithms, simulated annealing, Bayesian networks, and multi-agent systems (Marwala 2007, 2009, 2010, 2012; Marwala and Lagazio 2011). Some other approaches that were studied included game theory, control systems, Fourier transforms and wavelet transforms. The book introduced important themes such as economic data handling and modeling as well as prediction, knowledge discovery including data mining and causality versus correlation. It also outlined some of the common problems in economic modeling with regards to data handling, modeling and data interpretation. The book analyzed various economic data such the stock market, inflation, credit rating, option pricing, portfolio optimization and described important subjects such as inflation targeting.

The book introduced robust methods for economic data analysis and these were the mean, variance, kurtosis, fractals, frequency, time-frequency analysis techniques and stationarity. The Bayesian and the evidence frameworks were applied to generate an automatic relevance determination (ARD) tool. The ARD tool was used to evaluate the relevance of economic variables that were important for driving the consumer price index (CPI).

T. Marwala, *Economic Modeling Using Artificial Intelligence Methods*, Advanced Information and Knowledge Processing, DOI 10.1007/978-1-4471-5010-7_14, © Springer-Verlag London 2013

The multi-layered perceptron, radial basis functions and support vector machines were applied to model the CPI. The results indicated that the SVM gave the best results followed by the MLP and then the RBF.

Support vector machines and the multi-layered perceptron methods were applied using the Bayesian method to model American options and the results indicated that the MLP gave better results than the SVM. This book also introduced rough set theory and applied this to stock price prediction and observed high accuracy on classifying the daily movements of the Johannesburg Stock Exchange's All Share Index. The book applied auto-associative networks based multi-layered perceptron with genetic algorithms, particle swarm and simulated annealing optimization techniques for modeling manufacturing data and demonstrated that simulated annealing performed marginally better, followed by genetic algorithms and then the particle swarm optimization technique.

Furthermore, this book treated a predictive system as a missing data problem *i.e.* correlation machine and compared it to treating it as a cause and effect exercise *i.e.* causal machine. The correlation machine applied the autoassociative network, while the causal machine used the ARD. These approaches were applied to model the CPI and credit scoring. The ARD technique was found to be able to asses the causal relationships between the variables and the causal machine was found to perform better than the correlation machine for modeling credit scoring data while the correlation machine was found to perform better than the causal machine for modelling the CPI.

Genetic algorithms (GA) were applied for the continual rebalancing of portfolios. When both risk and return were targeted, the results showed that a GA was a viable tool for optimizing a targeted portfolio. The book applied an incremental learning procedure to predict the financial markets movement direction. Incremental learning was found to provide good results on adapting the weak networks into a strong learning algorithm that has confidence in all its decisions. The procedure was found to increase confidences in correctly classified instances and decrease confidences in misclassified instances after successive training sessions.

This book also simulated the stock market and implemented these within the game theory framework and the results indicated that this approach was a viable method for simulating the stock market. A control system approach was built for inflation targeting. The input/output model was built using a multi-layered perceptron network and a closed loop control strategy was adopted using GA to control inflation through the manipulation of interest rate. The calculated interest rate was compared to the historical inflation rate to evaluate the effectiveness of the control strategy and good results were obtained. The book also studied the role of trade in promoting peace and healthy economic activities, and the results indicated that trade was important for maintaining peace.

14.2   Further Work

For further studies, articial intelligence should be used to predict prices of crucial minerals, such as platinum, and relate these to economic growth of resource based economies. Another area for further study is the application of artificial intellence in remanufacturing, an area that is very important for modern industrialization. This book applied missing data approaches to classification and regression in economic modeling. For further studies, other techniques should be applied for missing data approaches to regression and classification, and these should include the Expectation Maximization Approach, Random Forrest, Firefly Algorithm, and Artificial Immune Systems.

References

Marwala T (2007) Computational intelligence for modelling complex systems. Research India Publications, New Delhi

Marwala T (2009) Computational intelligence for missing data imputation, estimation and management: knowledge optimization techniques. IGI Global Publications, New York

Marwala T (2010) Finite element model updating using computational intelligence techniques. Springer, London

Marwala T (2012) Condition monitoring using computational intelligence methods. Springer, London

Marwala T, Lagazio M (2011) Militarized conflict modeling using computational intelligence techniques. Springer, London

Biography

Tshilidzi Marwala, born in Venda (Limpopo, South Africa), is the Dean of Engineering at the University of Johannesburg. He was previously a Full Professor of Electrical Engineering, the Carl and Emily Fuchs Chair of Systems and Control Engineering, as well as the South Africa Research Chair of Systems Engineering at the University of the Witwatersrand. He is a professor extraordinaire at the University of Pretoria and is on boards of EOH (Pty) Ltd and Denel. He is a Fellow of the following institutions: the Council for Scientific and Industrial Research, South African Academy of Engineering, South African Academy of Science, TWAS – The Academy of Science of the Developing World and Mapungubwe Institute of Strategic Reflection. He is a senior member of the IEEE and distinguished member of the ACM. He is a trustee of the Bradlow Foundation as well as the Carl and Emily Fuchs Foundation. He is the youngest recipient of the Order of Mapungubwe and was awarded the President Award by the National Research Foundation. He holds a Bachelor of Science in Mechanical Engineering (Magna Cum Laude) from Case Western Reserve University, a Master of Engineering from the University of Pretoria, Ph.D. in Engineering from the University of Cambridge and completed a Program for Leadership Development at Harvard Business School. He was a post-doctoral research associate at Imperial College (London) and was a visiting fellow at Harvard University and Wolfson College (Cambridge). His research interests include the applications of computational intelligence to engineering, computer science, finance, social science and medicine. He has successfully supervised 39 masters and 9 Ph.D. students, published over 270 refereed papers and holds 3 patents. He is on the editorial board of the *International Journal of Systems Science* (Taylor & Francis) and his work has been featured in prestigious publications such as *New Scientist*. He has authored five books: *Condition Monitoring Using Computational Intelligence Methods* by Springer (2012), *Militarized Conflict Modeling Using Computational Intelligence Techniques* by Springer (2011); *Finite*

Element Model Updating Using Computational Intelligence Techniques by Springer (2010); *Computational Intelligence for Missing Data Imputation, Estimation and Management: Knowledge Optimization Techniques* by IGI Global Publications (2009); *Computational Intelligence for Modelling Complex Systems* by Research India Publications (2007); and he has a sixth book, *Economic Modeling Using Artificial Intelligence Methods* with Springer's computer science list.

Index

A

Acceptance probability function, 130, 131

Activation function, 51, 67–70, 76, 88, 96, 132, 138, 142, 144, 181, 186, 199, 200, 223, 227

Autoassociative network, 8, 15, 16, 122–123, 132, 133, 138, 139, 141, 253, 254

Average, 1, 23–26, 34, 35, 123, 133, 156, 157, 162, 165, 179, 186, 188, 204, 205, 208, 242, 244

B

Back-propagation method, 51–52, 89

Bagging, 175, 176, 201

Bayesian framework, 46, 54–55, 65, 72, 87, 223, 243, 247

Bayesian method, 54, 56, 92, 223, 254

Bayesian neural network, 87–89, 131, 246, 247

Black-Scholes model, 84–87, 94

Boolean reasoning (BR), 102, 109, 114, 115

Boosting, 175–177, 182, 188, 201

Boundary region, 108

BR. *See* Boolean reasoning (BR)

C

Causality, 6, 14–16, 30, 31, 137–151, 253

Committee of networks, 84, 195, 202, 203

Conjugate gradient method, 52–54, 58, 59, 68, 144, 148

Control system, 10–11, 16, 138, 159, 221, 225, 227, 235, 246, 253, 254

Cooling schedule, 132

Correlation, 6, 16, 35, 37, 46, 58, 92, 94, 119, 122, 137–151, 156, 217, 253, 254

Crossover, 126–128, 132, 142, 160, 161, 163, 196, 207, 208, 225–227

D

Decision rules, 58, 105

Derivatives, 2–3, 57, 85

E

Econometrics, 5–6, 16, 66

Economic development, 3–4

EFB. *See* Equal-frequency-bin partitioning (EFB)

Entropy based discretization, 110

Equal-frequency-bin partitioning (EFB), 102, 103, 109, 113–115

Evidence framework, 15, 45, 58, 60, 75, 245, 253

F

Fourier transform (FFT), 14, 15, 29–32, 253

Fractal dimension, 32, 33, 35

Frequency domain data, 14, 15, 24, 28–31

Fuzzy set, 103, 240–242

G

Game theory, 195–210, 225, 253, 254

Genetic algorithm, 5, 9–11, 16, 72, 103, 119, 121–129, 132, 133, 140–142, 144, 151, 155, 157, 159–163, 165, 168, 179, 182, 195, 196, 206–207, 223–227, 254

Granger causality, 15, 138, 145, 146

H
HMC. *See* Hybrid Monte Carlo (HMC)
Hurst, 24, 32–37, 40
Hybrid Monte Carlo (HMC), 87, 89–92, 96, 223, 246
Hyper-parameters, 46, 47, 56–58, 89, 146–149, 244, 245, 248

I
Incremental learning, 8–9, 174–175, 178–182, 188, 254
Indiscernibility relation, 102, 104–105
Industrialization, 2–4, 255
Inflation, 5, 23, 58, 65, 140, 215–228, 253, 254
Information
   system, 104, 105, 107–109
   table, 103, 105

K
Kurtosis, 14, 24, 27–28, 40, 87, 253

L
Learn++ incremental learning, 182–184
Linear assumption, 6

M
Markov chain Monte Carlo, 54, 95, 131, 223, 244
Maximum likelihood technique, 227
Metropolis algorithm, 91, 95, 130, 131, 244
MLPs. *See* Multi-layer perceptrons (MLPs)
Model selection, 54, 69, 71, 84, 242
Modern portfolio theory, 157–158, 166
Monte Carlo, 54, 57, 84, 86, 89, 94–96, 129–131, 201, 223, 246
Multi-agent system, 9, 16, 197–199, 206, 207, 253
Multi-layer perceptrons (MLPs), 48, 49, 60, 67–68, 84, 87, 119, 139, 144, 147, 179, 185–186, 188, 198, 200, 202, 222–224, 226, 245, 253
Multi-scale fractal dimension, 32
Mutation, 126, 128

N
Naïve algorithm, 110
Network weights, 49–52, 56, 57, 69–71, 90, 91, 223, 227, 244, 247

Neural network, 5, 23, 46, 66, 83, 102, 119, 137, 160, 174, 198, 217, 240
Neuro-rough set, 243–245, 247

O
Objective function, 45, 49–53, 58, 121, 123, 129, 131, 132, 160, 226, 239, 240, 247
Options, 2–3, 15, 83–87, 94, 95, 97, 254

P
Particle swarm optimization, 10, 11, 16, 68, 84, 119, 121, 123–125, 133, 202, 253, 254
Political stability, 2, 4–5
Position, 89–91, 109, 123–125, 198, 233, 249
Posterior distribution function, 244
Principal component analysis, 46
Prior distribution function, 88
Prior probability distribution, 55

R
Radial basis function (RBF), 7, 15, 48, 65, 67–71, 73, 78, 93, 96, 198–200, 202, 240, 254
Reduct, 107–108, 111, 113, 115, 243
Rescaled range, 32–35, 180
Rough set, 8, 15, 101–105, 107, 111–113, 115, 126, 243, 244, 247

S
Scaled conjugate gradient, 52–54, 58–60, 68, 71, 121, 122, 144, 148, 200, 227, 245
Selection, 8, 10, 24, 54, 69, 71, 72, 84, 122, 126, 128, 130, 132, 157, 160–163, 175, 179, 180, 188, 201, 207, 226, 242
Set approximation, 107
Simulated annealing, 89, 103, 121–123, 127, 129–130, 132, 133, 223, 253, 254
Stacking, 175, 176, 201, 202
State space, 89, 91, 92, 95, 130, 131
Stationarity, 24, 36–40, 218, 253
Stock market, 2, 15, 16, 26, 31, 32, 35, 87, 110, 112–115, 126, 145, 173, 180, 195–210, 253, 254
Support vector machines, 46, 72, 85, 144, 247

T
Takagi-Sugeno neuro-fuzzy model, 241, 242
Time domain data, 14, 24–28
Time-frequency data, 14, 15, 24, 31, 253

Transformer bushing, 179
Transition probabilities, 130

V
Variance, 14, 15, 23, 24, 26–27, 36, 38–40,
        70, 71, 119, 130, 145, 146, 156–158,

165, 166, 176, 179, 201, 217–220, 242,
244, 253

W
Wavelet transform, 15, 31, 33

Printed in the United States
By Bookmasters